ROUTLEDGE LIBRARY EDITIONS: THE ECONOMY OF THE MIDDLE EAST

Volume 23

OIL, INDUSTRIALIZATION & DEVELOPMENT IN THE ARAB GULF STATES

OIL, INDUSTRIALIZATION & DEVELOPMENT IN THE ARAB GULF STATES

ATIF A. KUBURSI

Taylor & Francis Group
LONDON AND NEW YORK

First published in 1984

This edition first published in 2015
by Routledge
2 Park Square, Milton Park, Abingdon, Oxon, OX14 4RN

and by Routledge
711 Third Avenue, New York, NY 10017

Routledge is an imprint of the Taylor & Francis Group, an informa business

© 1984 Atif Kubursi

All rights reserved. No part of this book may be reprinted or reproduced or utilised in any form or by any electronic, mechanical, or other means, now known or hereafter invented, including photocopying and recording, or in any information storage or retrieval system, without permission in writing from the publishers.

Trademark notice: Product or corporate names may be trademarks or registered trademarks, and are used only for identification and explanation without intent to infringe.

British Library Cataloguing in Publication Data
A catalogue record for this book is available from the British Library

ISBN: 978-1-138-78710-0 (Set)
eISBN: 978-1-315-74408-7 (Set)
ISBN: 978-1-138-81015-0 (Volume 23)
eISBN: 978-1-315-74471-1 (Volume 23)
Pb ISBN: 978-1-138-82026-5 (Volume 23)

Publisher's Note
The publisher has gone to great lengths to ensure the quality of this reprint but points out that some imperfections in the original copies may be apparent.

Disclaimer
The publisher has made every effort to trace copyright holders and would welcome correspondence from those they have been unable to trace.

OIL, INDUSTRIALIZATION & DEVELOPMENT IN THE ARAB GULF STATES

ATIF A. KUBURSI

CROOM HELM
London • Sydney • Dover, New Hampshire

© 1984 Atif Kubursi
Croom Helm Ltd, Provident House, Burrell Row,
Beckenham, Kent BR3 1AT
Croom Helm Australia Pty Ltd, First Floor, 139 King Street,
Sydney, NSW 2001, Australia

British Library Cataloguing in Publication Data

Kubursi, Atif
   Oil, industrialization and development
   in the Arab Gulf States.
   1. Persian Gulf States – Economic
   conditions
   I. Title
   330.953'6053   HC415.3

ISBN 0-7099-1566-7

Croom Helm, 51 Washington Street, Dover,
New Hampshire 03820, USA

Library of Congress Cataloging in Publication Data

Kubursi, A.A.
   Oil, industrialization, and development in the
Arab Gulf states.

   Bibliography: p.
   Includes index.
   1. Persian Gulf States – Economic conditions.
2. Petroleum industry and trade – Persian Gulf
States. I. Title.
HC415.3.K83   1985   330.953'6   84-19988
ISBN 0-7099-1566-7

Typeset by Mayhew Typesetting, Bristol, UK
Printed and bound in Great Britain
by Billing & Sons Limited, Worcester.

CONTENTS

List of Tables and Figures

Acknowledgements

1. Introduction	1
2. The Economies of the GCC Region: The Macro Perspective	7
3. Oil and Gas	41
4. The Non-oil Sectors: The Quest for Diversification	71
5. Agricultural Development in the GCC Region: Is It Possible?	75
6. Industrialization: Prospects and Problems	101
7. Conclusion	131
Notes	137
Index	139

TABLES AND FIGURES

Tables

2.1	Gross Domestic Product of the GCC Region, by Sector of Origin and Country, 1971	23
2.2	Gross Domestic Product of the GCC Region, by Sector of Origin and Country, 1976	24
2.3	Gross Domestic Product of the GCC Region, by Sector of Origin and Country, 1981	25
2.4	Index Numbers of Gross Domestic Product of the GCC Region at Constant Prices, by Sector of Origin	26
2.5	Expenditures on Gross Domestic Product, 1971	27
2.6	Expenditures on Gross Domestic Product, 1976	27
2.7	Expenditures on Gross Domestic Product, 1981	28
2.8	Gross Domestic Product of the GCC Region, by Sector of Origin and Country: Percentage Distributions Among Countries, 1971	29
2.9	Gross Domestic Product of the GCC Region, by Sector of Origin and Country: Percentage Distributions Among Countries, 1976	30
2.10	Gross Domestic Product of the GCC Region, by Sector of Origin and Country: Percentage Distributions Among Countries, 1981	31
2.11	Commodity Trade Within the GCC Region and With the Rest of the World, 1981	32
2.12	Exports from the GCC Region, Total and Percentage Distribution Among Countries, 1975-1981	33
2.13	Imports to the GCC Region, Total and Percentage Distribution Among Countries, 1975-1981	33
2.14	Exports from the GCC Region to Industrial and Developing Countries, Percentage Shares, 1975-1981	34
2.15	Imports to the GCC Region from Industrial and Developing Countries, Percentage Shares, 1975-1981	35
2.16	Balance of Payments of the GCC Member Countries, 1979-1981	36
2.17	Size and Selected Characteristics of the Population of the GCC Region, by Country, Various Years	37

Tables and Figures

2.18	Size and Selected Characteristics of the Labour Force of the GCC Region, by Country, 1981	38
2.19	Resource Base and Related Environmental Features of the GCC Region: An Overview	39
3.1	GCC Proven Crude Oil Reserves, 1970-1980	53
3.2	GCC Crude Oil Production, 1970-1980	54
3.3	GCC Crude Oil Exports, 1970-1980	55
3.4	Production and Utilization of Natural Gas in the GCC Region: Selected Years	56
3.5	GCC Gross Production of Natural Gas, 1971-1980	57
3.6	GCC Proven Natural Gas Reserves, 1971-1980	58
3.7	Average Chemical Structure of Associated Gas in Selected GCC Countries	59
3.8	Average Chemical Structure of Natural Gas (Dry) in Selected GCC Countries	60
3.9	Oil Refining Capacity in the Arabian Gulf	61
3.10	Oil Refineries in the GCC Region, 1980	62
3.11	Liquid Natural Gas Projects in the GCC Region	63
3.12	Existing and Planned Fertilizer Industries in the GCC Region	65
3.13	Existing and Planned Petrochemical Projects in the GCC Region	67
5.1	Land Area in the Countries of the GCC Region	87
5.2	Total Population, Agricultural Population and Population Economically Active in Agriculture, in Thousands, for Selected Years, 1970-1980	90
5.3	Domestic Agricultural Production in the GCC, 1970-1981	93
5.4	Principal Tree Crop Production, Selected Years	94
5.5	Principal Cereal Crops, Area Planted, Yield and Production in Oman and Saudi Arabia, Selected Years	95
5.6	Selected Fruit and Vegetable Production Statistics in the GCC Countries, Selected Years	96
5.7	Numbers of Livestock in the GCC Countries, 1976, 1977 and 1978	98
5.8	Agricultural and Total Development Expenditures by the GCC Countries, 1970-1980	99
6.1	Consumption of Iron and Steel, 1978	122
6.2	Arab Gulf: Profile of Steel Industry, 1982	123
6.3	Comparison of Investment and Operating Costs Per Tonne	

	of Aluminium, 1980 Prices	124
6.4	Consumption of Semi-finished and Finished Aluminium	125
6.5	Arab Gulf: Cement Production Capabilities	125
6.6	Selected Licensed Food Processing Factories in Saudi Arabia, 1980	126
6.7	Projected Average Annual Demand for Telephone Exchange Lines	127
6.8	Projected Average Annual Demand for Telephone Instruments	127
6.9	Projected Average Annual Demand for Telex Exchange Lines and Telex Machines	128
6.10	Projected Average Annual Demand for Installed Generation Capacity	128
6.11	Projected Average Annual Demand for Transformers	129

Figure

5.1	Agricultural Sector Linkages	86

*Dedicated to
Jinan, Marwan and Omar*

ACKNOWLEDGEMENTS

I would like to express my thanks to the Petroleum Information Commitee of the Arab Gulf States for their financial support which enabled me to undertake this project, and the complete freedom I enjoyed in writing it.

I owe a great debt to many of my colleagues and friends who have contributed, each in a special way, to the development and progress of this study. My greatest debts are to Dr Walid Sharif who first conceived this project, and to my colleagues at McMaster University and abroad who worked with me on UNIDO's project on 'The Resource Base For Industrialization in the Gulf Co-operation Council Countries: A Framework For Co-operation', David Butterfield, Kenneth Chan, Frank Denton, David Freshwater, Najim Kassab, Byron Spencer and Marios Tsezos. Special thanks are also due to Herman Muegge and Se Hark Park at UNIDO, Ribhi Abu Haj and George Haddad at ECWA, Abdullah El-Kuwaiz, Abdul-Rahman Ohali, Abdul-Rahman Al-Muhana and Nabeel Akeel of the Gulf Co-operation Council and to Betty May Lamb for her excellent typing and research.

I alone bear the full responsibility for the views expressed and the judgements made in this book.

1 INTRODUCTION

1.1 The Options

Were oil supplies everlasting, and the demand for oil strong and continuous, economic diversification would be pointless. The governments of the region would instead need only to ensure the distribution of oil revenues among the population. However, this is far from being the case in the Gulf region.[1] Oil reserves are finite and non-renewable, and the world demand for oil from the region is simply not stable. At recent rates of utilization oil in the GCC (Gulf Co-operation Council) region will run out in the lifetime of the present generation (Bahrain, Qatar and Oman), its children (UAE), or its grandchildren (Kuwait and Saudi Arabia).

Culturally, historically, geographically and politically the GCC states represent a rare instance of regional homogeneity, and the problems that confront them are similar also. Collectively and individually they face the striking realization that, unless priorities and plans are set with care, the gestation period of economic development may be longer than the expected life of their hydrocarbon resources. They face the historical challenge of accumulating enough productive capital (human and physical) in the non-oil sectors of their economies, and of raising productivity sufficiently in those sectors, to offset the drawing down of oil reserves. In a sense, they are in a race against time. Standards of living in the region have risen markedly, of course. Extensive welfare programmes have been introduced which are a source of admiration by even the most ardent socialists, and there have been massive improvements of infrastructure, of the educational system, and even of heavy industry. However, this has been accomplished at the cost of substantial capital consumption, in the sense that oil reserves have been depleted at much faster rates than human and physical capital have been created to replace them.

1.2 Development Through Co-operation

The realization that the region has not yet reached the threshold of sustained economic growth independent of oil revenues is coupled with the realization that the necessary rate of sustained growth may be possible only within a framework of regional co-operation and economic integration. The general perception of vulnerability associated with the vast wealth of the region, its limited military capabilities, its relatively small and dispersed population, the extensive land mass of the Arabian Peninsula and the neighbouring Gulf war, has added urgency to the quest of co-operation.

Political boundaries in the GCC region, the much larger Arab region or, for that matter, the Third World as a whole, do not necessarily correspond with optimum geographic boundaries from an economic point of view. Thus the development of closer economic contacts among these countries may be of significant mutual benefit.

Increasing trade among the member countries of the GCC is not the only objective of their regional economic association. As will be demonstrated later, their intra-regional trade flows represent only a small fraction of their total trade flows (less than 4 per cent). The member countries are, at present, too similar for much trade to take place. Indeed, the end products of each are currently competitive rather than complementary with those of others in the region. Thus the exportable portion of each country's output must find its market outside the immediate region while a wide range of commodities must come from sources entirely outside the region, or be done without.

Little economic complementarity prevails within a region comprised entirely of developing countries, such as those in the GCC region, and thus the basis for mutually advantageous trade tends to be weak. But complementarity can become a fact in the future. With the emergence of industrial activity within the region, greater complementarity will likely result and this would encourage a larger volume of intra-regional trade. At the same time, the introduction of new industrial activities may be long delayed, or even found to be impossible, unless a large enough market can be assured for the end products. There is, therefore, a strong case for the strengthening of regional trade arrangements — not to yield immediate gains from the exchange of goods available in the prevailing economic circumstances, but rather as a means to promote and accelerate economic development itself. Then, as economic development proceeds, and member countries achieve higher levels of economic activity generally, they will find themselves engaged in more extensive intra-regional trade. The small economic size of most countries of the GCC region has restricted their independent economic development. But with the extension of the boundary of production to include the whole GCC region, industrialization opportunities will increase and this, in turn, will augment the possibilities for increased specialization and trade. Put simply, the regional market, by making intra-regional trade possible, may help to encourage new activities which would, in turn, increase intra-regional trade. This is a dynamic process that feeds on itself; once put in motion, it generates its own momentum.

1.3 The Strategy

The options for the GCC governments are restricted but clear. The GCC countries have but a limited time in which to make use of oil revenues to create a viable economy that will sustain a relatively high level of income after the oil era. Such an economy must be achieved through a long-term programme of maintaining a high ratio of investment to non-oil GDP while at the same time sustaining a major effort to enhance the skill levels of the domestic labour force and to widen and deepen co-ordinated regional development.

High ratios of investment to non-oil GDP in the GCC countries are possible with little if any sacrifice of present consumption. Savings have not been generated by forgoing consumption, but have accrued principally from the oil sector in such amounts that the funds available for investment have, until recently, exceeded the physical and economic possibilities for utilizing them productively. This is the well-known constraint of limited absorptive capacity. But this constraint is not exogenous and fixed. It may be relaxed through the augmentation of markets, through inter-country policy harmonization, by dismantling the domestic constraints and by extending the regional boundaries of production within the GCC area and beyond it. Absorptive capacity may also be extended by a programme of co-ordinated investments, thereby taking advantage of backward and forward linkages in production sequences. In short, absorptive capacity of the regional economy can be deliberately extended and augmented.

The accumulation of large financial surpluses in the early 1970s by the GCC countries preceded any deliberate plan for their domestic absorption or investment abroad. There was no historical experience on which to base predictions of future growth in domestic absorption and, therefore, no reason to question the economic rationality of a surplus of the magnitude realized. However, these countries are no longer oblivious to the risks of accumulating fixed-income-yielding assets in a basically inflationary world. They are no longer satisfied with their role as just the residual suppliers of the world crude oil requirements. They have accommodated the industrialized countries' oil demands almost to the detriment of their own interests with little or no economic or political quid pro quos. Now they are bent on the creation of an advanced and integrated economy. This is not an easy task as they still face inordinate obstacles: harsh climates; lack of sufficient arable land, water, minerals; a relatively high degree of illiteracy; a scarcity of managerial talent or experience; at best, a very small skilled labour force and, likewise, a shortage of available unskilled labour. Even in the best of circumstances the rapid increase in oil revenues was bound to create a great deal of confusion, mismanagement and substantial duplication of effort. Now it is ten years since the first substantial correction of oil prices, past rationalization of failures is no longer admissable particularly since the emergence of the GCC on 4 February 1981 and the ratification of the Economic Agreement on 8 June 1981.

1.4 The Economic Agreement

Although the GCC is not exclusively an economic association, it is perhaps the most promising effort in economic co-operation among developing countries in recent time. In fact, the GCC has been successful in establishing a number of agreements in the fields of defence, foreign affairs, education and energy. However, the Economic Agreement (EA) is perhaps the GCC's most important achievement. The EA is embodied in a comprehensive document which details the framework, mechanisms and principles of co-ordination, harmonization and integration of economic activity in the region.

The EA consists of seven parts and 28 articles. The major elements of the agreement are the following:

(1) a free-trade area in the region with no tariffs on regional products and a common tariff on non-regional outputs;
(2) a pooling of bargaining power when dealing with trading partners outside the region;
(3) a common market area in which citizens of the Council member countries are free to move, work, own, inherit and bequest within each and every country of the Council;
(4) a call for harmonization of development plans with the aim of complete integration;
(5) a common oil policy and a co-ordinated policy of industrialization based on oil resources;

(6) a co-ordinated industrial development policy for the region as a whole;
(7) special emphasis on establishing and promoting joint projects in all sectors with the aim of tying the production chains of the region into a common development sequence;
(8) co-operation in the development of local technology and the joint acquisition of foreign technology;
(9) pooling of human resources to prevent harmful competition for scarce labour;
(10) linking the regional markets through road, rail, air and water transportation networks;
(11) a common legal framework for trade and investment in the region; and
(12) a common development aid policy.

There is a wide range of possible instruments or patterns for economic co-operation among countries, extending all the way from customs unions to full integration. The GCC's Economic Agreement seems to have spanned much of this range. Another distinctive feature is the adoption of the principle of gradualism, so that co-operation is expected to proceed pragmatically and cumulatively, rather than abruptly. Furthermore, and perhaps uniquely, the EA stipulates an important role for the private sector in the implementation and maintenance of economic interrelationships.

The framework of co-operation is a dynamic one. The emphasis on the co-ordination of investments, on development effort and on complementary activities takes its context from a desire to foster growth and development by reaping the benefits of rationalization of production and intra-regional specialization. In this regard, the EA is balanced, realistic and forward looking.

It is clear that the EA recognizes the formidable barriers to market integration imposed by the physical environment of the region. Thus markets and production activities are to be integrated by a deliberate linking of road, rail and air transport networks. The economic 'balkanization' of the area can be eliminated only as national infrastructures come to be joined and integrated within a common regional framework, and this has come to be recognized.

The common characteristic of an overwhelming and continuing dependence on oil may be related, in some measure, to the small size of the individual economies of the GCC member states. This limited size may have inhibited economic diversification of the national economies in the past. Collectively, though, the countries of the GCC may be able to develop and sustain a much more diversified economic base. Not only is the market greater, but so are the pools of natural resources and the bargaining power of the countries when they act in concert.

The purpose of this study is to explore the development potential of the region within a framework of co-operation among the GCC countries and within the broader context of Arab development. The basic contention of the study is that economic development in the GCC region is inseparable from economic development in the Arab world at large. The two phenomena are intertwined and complementary and one cannot be sustained without the other; the two reinforce one another.

1.5 The Plan of the Book

A brief outline of the text is presented here to set the tone for the development of our conclusions. The next chapter provides a macroeconomic perspective of the Arab Gulf states and the region's economy as a whole. Chapter 3 deals with oil and gas and downstream activities in this sector. Chapter 4 considers the prospects and problems of diversification with limited water and labour resources. Two sectors are singled out as promising for further development. Agricultural activity and developments within this sector are the subject of Chapter 5, whereas industrialization is the subject of Chapter 6. The book concludes with the formulation of a development strategy for the region.

2 THE ECONOMIES OF THE GCC REGION: THE MACRO PERSPECTIVE

2.1 Introduction

This chapter is concerned with the overall economy of the GCC region and with differences, similarities and relationships among the individual economies of its six members. The aim of the chapter is to provide a broad picture of relevant economic and associated demographic and environmental characteristics that will serve as a background to the more detailed analysis in subsequent chapters. We begin by providing a synoptic view of the region's production patterns and the industrial composition of its gross domestic product (GDP) and the allocation of the product among major expenditure categories, as reflected in national accounting estimates. Particular attention is given to the role of oil exports as the driving force behind economic expansion and to the trading relationships of the GCC countries, both among themselves and with the rest of the world. This is followed by a general overview of the region's natural resource base and its human resources.

2.2 The Structure of Production

Basic national accounting data relating to production and expenditures are provided in Tables 2.1-2.10 (see pp. 23-31). Tables 2.1-2.3 provide estimates of GDP, by country and sector of origin, for the years 1971, 1976 and 1981. The GDP figures on which these tables are based are expressed in US dollars for comparability and are at current prices. Table 2.4 provides indexes of GDP at constant 1970 prices. Finally, Tables 2.5-2.7 provide estimates of expenditure on GDP by final demand category (consumption, investment, exports and imports) and country, in current US dollars, whereas Tables 2.8-2.10 provide the percentage distribution of GDP among GCC countries in each sector.

We turn first to Table 2.3 and note that the total of the GCC region is estimated to have been about US$210 billion in 1981, expressed at purchasers' prices. The phenomenal growth in the value of the region's output is indicated by the corresponding totals for 1971 and 1976

(Tables 2.1 and 2.2): in 1971, the value of the GDP was only about US$11 billion; by 1976, it had risen to about US$79 billion, an increase of more than six-fold; and by 1981, it had reached the US$210 billion level just noted, or about 19 times the level of a decade earlier. The driving force behind the growth in the value of production was, of course, the sharp rise in crude oil prices in the 1970s and, to a much lesser degree, the overall increases in the annual volume of oil extracted within the region. That the massive increase in the region's oil revenues induced very rapid expansion in the economy at large is evidenced by a 14-fold increase in the value of manufacturing production between 1971 and 1981, a 15-fold increase in the service-producing sector and a 42-fold increase in construction.

The population was increasing quite rapidly too over this period. Nevertheless, the value of GDP *per capita* is estimated to have risen from about US$1,300 in 1971, to a little under US$17,000 in 1981. This increase — and the associated increase in the aggregate GDP — should be kept in proper perspective. Much of the gain was offset by inflation — by higher prices of consumption and investment goods to purchasers in the GCC countries — but the remaining very large real increase should be interpreted with care as well. To the extent that it was the result of selling off some of the region's oil supply in world markets, or of consuming oil domestically, the increase might better be viewed as a conversion of wealth: oil under the ground may be regarded as a very large component of the region's national wealth, and the extraction of it implies the using up or transformation of this form into some other form of wealth, or into current consumption. In this sense, the annual income or production figures based on conventional national accounting procedures may be misleading; they may overstate the levels of income in a fundamental sense — in the sense, that is, of sustainable levels of income.

The rise in crude oil prices in the 1970s might thus be regarded as a *wealth* effect, rather than as a 'true' shift in income level. The price rise implied a drastic upward revaluation of the GCC region's underground resources — its stock of natural capital — if one uses current prices to do the valuation. But in a deeper sense, the present value of the underground wealth depends not just on current prices but on expectations about future prices, and these, in turn, depend on expectations about future world demand and supply conditions, including the effects of technological change and price-induced energy substitution. For present purposes, the main point to be emphasized is that conventional national accounting measures (such as those in our tables) may overstate the national income by including what might better be regarded as wealth conversion than as a flow of income or production that is necessarily sustainable into the long-term future. It would seem highly desirable to develop a set of national accounts for the region in which allowance is made for resource depletion. This is a difficult task and can yield only rough approximations. Nevertheless, even crude estimates which incorporate such an allowance would be more realistic than the conventional ones, which do not.[2]

The dominance of oil is immediately evident from the high proportions of GDP originating in the sector labelled 'mining, quarrying and oil extraction'. In 1971, about 61 per cent of the region's GDP, at current prices, came from this sector; by 1976, the proportion had risen to about 65 per cent; in 1981, at 63 per cent, it had fallen a little, but was still higher than a decade earlier. In five of the six member countries, the proportions were either higher in 1981 than in 1971, or unchanged. The exception was Bahrain, where there had been a very sharp drop — from about 72 per cent in 1971 to 37 per cent in 1976, and then to 33 per cent by 1981 — as the result of a marked reduction in the rate of oil extraction.

An examination of the 1981 country distributions indicates a remarkable feature: with one exception (again Bahrain) the intra-country sector distributions look quite similar, in very broad terms. The proportions of GDP value originating in 'mining, quarrying and oil extraction' for Kuwait, Oman, Qatar, Saudi Arabia and the UAE all lie between 62.7 and 69.2 per cent; the proportions in all commodity-producing sectors combined lie in an even narrower range — from 76.9 to 79.3 per cent; and the proportions in all service-producing sectors combined lie between 20.7 and 23.1 per cent. Given the realiability of the data, one could almost say (in this very broad sense) that the proportionate distribution shown for any one of the five countries would serve as an estimate for any one of the others. The similarity is the more remarkable when one recalls the pronounced differences in population size, rural—urban distributions, and other demographic or labour-related characteristics (to be discussed in the following sections).

As one might expect, the differences from one country to another are seen to be greater when one looks at Table 2.3 in more detail. Bahrain and Kuwait are estimated to have had the largest proportions of GDP value originating in manufacturing in 1981 — 9.5 per cent in the first and 6.1 in the second — while the lowest proportion, 1.0 per cent, is recorded for Oman. Agriculture ranges from 0.2 per cent in Kuwait to 2.1 per cent in Bahrain. Construction ranges from 2.7 per cent in Kuwait to 11.0 in Saudi Arabia, and to 13.2 per cent in Bahrain. Other sectors show similar kinds of variation.

Percentage distributions among countries for each production sector are shown in Table 2.8 to 2.10. The dominant size of Saudi Arabia in 1981 is reflected in the distribution of total GDP, and in all of the sector distributions as well: the Saudi Arabian economy produced 65 per cent of the region's GDP in that year, about 65 per cent of its crude oil, some 65 per cent of its manufacturing output, and so on. The UAE accounted for about 15 per cent of the total GDP, and Kuwait for 12 per cent.

A comparison of the distributions for 1981 (Table 2.10) with those for 1971 (Table 2.8) is revealing, and indicates again the significance of oil production for the interpretation of the conventional GDP measures. In 1971, Kuwait and Saudi Arabia each produced about two-fifths of the value of crude oil output of the GCC region. Kuwait's share of the total GDP was about 35 per cent in 1971, and Saudi Arabia's about 46 per cent. By 1981, Kuwait had scaled down its level of oil production very sharply in order to lengthen the life of its underground reserves; as noted, its share of total GDP had, therefore fallen to only 12 per cent, while Saudi Arabia's had increased to 65 per cent. This highlights again the need for care in interpreting the conventional national accounting income and production figures; the marked shift in the relative shares of the two countries was much more the result of differences in the rates of conversion of their natural underground wealth than of fundamental changes in their abilities to generate sustainable flows of income.

Indexes of the region's GDP, in real or constant-dollar terms, are shown by sector in Table 2.4 for the period 1960 to 1979, together with annual growth rates for selected intervals within this period; the indexes have base 100.0 in 1970. Overall, the real GDP increased by 479 per cent between 1960 and 1979, and by 149 per cent between 1970 and 1979. These increases are far less than those in the current-dollar levels discussed above, but still large. They imply an annual rate of growth of 9.7 per cent over the entire 1960 to 1979 period, and of 10.7 per cent over the 1970 to 1979 period. Such rates are very rapid by any standard.

The increases in real production of crude oil are implicit in the indexes for mining, quarrying and oil extraction, since oil accounts for almost all of the output of this sector. From 1970 to 1979, the sector index rose by 89 per cent, or at a rate of 7.3 per cent per annum. Next to agriculture, fishing and trapping, this is the smallest rate shown for any sector in the table: manufacturing increased at 10.6 per cent per annum over the same period, electricity, gas and water at 15.3 per cent, construction at 16.1 per cent, wholesale and retail trade at 15.7 per cent, and transportation and communication at 17.7 per cent. An annual rate of increase of the order of 7 per cent in oil extraction is still high, of course, and it has important implications for the pace at which the known underground petroleum reserves are being exhausted. However, it is far less than the rate of growth of production value, since the latter reflects the very sharp price increases that occurred in the 1970s, as well as the lesser increases in physical volume.

The disposition of the GDP as evidenced by the national expenditure accounts of the GCC region and its member countries, is recorded in Tables 2.5-2.7, again for the years 1971, 1976 and 1981.

Bahrain has had annual deficits on its international trade account — large ones, relative to its GDP in 1971 and 1976, though quite small in 1981 — as these tables indicate. However, in all of the other countries, and in the region as a whole, there have been huge surpluses, reflecting the revenues generated by shipments of oil exports at sharply increased world prices. In 1981, the balance-of-trade surplus for the GCC region as a whole was equal to about 35 per cent of the region's total value of GDP; for the five individual surplus countries, the percentages ranged from 33 (in Saudi Arabia) to 68 (in Qatar). It is, of course, these massive trade surpluses that have provided the funds for large-scale investment and rapid economic expansion in the GCC region. Indeed, a major problem within the region has been (and continues to be) the management of the huge volume of capital funds generated by oil exports, and the rate at which the economies of the region can effectively absorb these funds in the form of investments in new plant and equipment, urban construction developments, improved economic infrastructure, and so on, or in the form of enhanced levels of consumption by the population at large. The inability to absorb the funds at anything like the rates at which they have accumulated has led to holdings of financial assets abroad of unprecedented magnitude, and these, in turn, have generated new investment revenue, and hence further accumulation of foreign holdings.

That the economies of the GCC countries are overwhelmingly dependent on foreign trade is a point that scarcely needs emphasis. However, the statistical evidence in the national accounts tables should at least be noted. For the region as a whole, exports represented some 72 per cent of the total GDP in 1981, and imports some 37 per cent.

In Bahrain, exports represented 120 per cent of the GDP, and imports 121 per cent. (These figures reflect the substantial re-export component of imports into Bahrain.) Of the remaining five countries, Saudi Arabia had the lowest ratio of exports to GDP, at 68 per cent, and Qatar had the highest, at 91 per cent. Imports in 1981 represented about 23 per cent of GDP in Qatar; in the other countries the proportion was in no case below 32 per cent.

Investment, in the sense of physical capital creation, absorbed the equivalent of 24.5 per cent of the total GCC gross domestic product in 1981; about 70 per cent of all such investment within the region took place in Saudi Arabia, while the United Arab Emirates accounted for about 16 per cent, and Kuwait for about 8 per cent.

The trading relations of the GCC countries are predominantly with the non-GCC world; flows of imports and exports within the region itself are generally small. Bahrain represents somewhat of an exception, inasmuch as it imports a substantial volume of oil from Saudi Arabia and has served as a point of transshipment for goods from abroad that are destined for other GCC countries, most notably to the United Arab Emirates. In the main, though, there is comparatively little direct economic interconnection among the GCC countries at the present time. This is a point of considerable importance since much closer trading relationships might have to be developed among the member countries if a co-ordinated programme of industrialization and intra-regional specialization in production were to be initiated by the GCC. The existing patterns of trade are examined in detail in the next section.

2.3 Intra-regional and External Economic Relationships

The contacts of the GCC countries with the rest of the world have increased substantially in recent years, and their economic relationships with other countries have become much more prominent and extensive. The main reason for this, of course, has been the vast increase in their wealth and in their strategic importance, as the world has become increasingly dependent on their oil and trade.

In section 2.3.1, the focus is on trade in commodities. Based on available data, classified according to the Standard International Trade Classification system (SITC), we will attempt to identify the nature and magnitude of commodity trade within the region, and between the region and the rest of the world. Special attention will be paid to the trade between the GCC countries and other Arab countries, as well as trade with the major industrial countries of the world and with non-Arab developing countries.

2.3.1 Commodity Trade

As has already been indicated, trade between countries in the GCC region is very limited in scope, and involves mostly the re-export of goods imported from abroad. The reason is simple: the countries are basically very similar; they extract and export their reserves of oil wealth and import almost everything else. Hence the possibility of trade within the region, other than that involving re-exports, depends on the development of domestic industries. While it is likely that such trade has increased in recent years as the national economies have become somewhat more diversified, it remains the case that trade with the rest of the world overwhelmingly dominates trade within the region.

We consider first the extent and nature of the trade flows involving the GCC countries, the types of commodities traded, and their countries of origin and destination. The discussion here is based largely on data obtained from two related sources: the IMF publication, *Direction of Trade Statistics Yearbook, 1982* (hereafter, DOTS Yearbook) and the UNIDO data tapes, *COMTRADE*. The first source provides data on the distribution by trade partners of total merchandise exports and imports for 154 countries for the years 1975 to 1981, while the second provides similar data at a commodity level of detail, based on SITC.

We start by considering the overall commodity trade flows, as reported in the DOTS Yearbook. Information relating to the 1981 total commodity flows within the GCC region and between each member country and the rest of the world as a whole is provided by Table 2.11. Measures of trade flows can be based on either export data or import data, or some combination of the two. Table 2.11 is based mainly on the import data for each of the member countries, along with each country's total reported exports. Figures for imports from the exports to the 'rest of the world' are calculated residually. As is clear in Table 2.11 the trade flows involving each country with others in the region are far smaller than those with nations outside the region. The only exception is Bahrain, which receives a large fraction of its imports from Saudi Arabia, mostly in the form of crude petroleum for refining and other materials for use in aluminium smelting, and a large fraction of whose exports are re-exports, going especially to the UAE. In all other cases, the exports to other GCC countries represent less than 4 per cent of total exports in 1981. On the import side, Saudi Arabia received about 1.2 per cent of its imports from other GCC countries, and Kuwait 3.2 per cent. The figures for the other countries are somewhat higher, including especially Bahrain, as already noted. Even though suitable data are not available on re-exports, it seems likely that, aside from the special link between Saudi Arabia and Bahrain, the bulk of the trade within the GCC involves re-exports of goods brought in from abroad.

Further summary information on the total volume of exports emanating from the GCC region over the period 1975 to 1981 is provided in Table 2.12. In this table are reported the total current dollar value of exports from the region (measured exclusive of exports to other GCC countries), and the percentage distribution of those exports by country of origin. There was a spectacular growth in the current value of exports, especially towards the end of the period, reflecting, in large part, the rise in oil prices which took place at that time. It is once again evident that Saudi Arabia dominates, accounting for about 60 per cent of total exports in each year between 1975 and 1979, more than 67 per cent in 1980 and over 71 per cent in 1981. The recent rise in its share reflects Saudi Arabia's decision to continue to increase oil production at a time when some of the other countries, most notably Kuwait, were cutting back.

A similar tabulation showing total imports and their percentage distribution within the GCC is provided in Table 2.13. The total value of imports also increased very rapidly over the period 1975 to 1981, following the increase in exports, but the value of commodity imports remained well below that of commodity exports. In fact, by 1980, and again in 1981, the commodity exports of the GCC region were estimated to exceed its imports by more than US$100 billion. Saudi Arabia has been the largest importer, and its proportion of the total has grown over the period, presumably reflecting in part the opening of extensive new port facilities of its own, and hence the reduction in the volume of imports received as re-exports from other GCC countries. The growth also reflects a relatively more ambitious attempt at rapid economic development.

The percentage shares of exports going to industrial and developing nations are shown in Table 2.14, for each country in each year from 1975 to 1981. Within the group of developing countries, we show separately the non-GCC Arab region. The bulk of exports — about

The Economies of the GCC Region: The Macro Perspective 15

70 per cent — goes to the industrial nations and most of the rest goes to the developing nations. (In addition, a small amount goes to the USSR and Eastern Europe, and to countries not specified.) The non-GCC Arab countries received about 2 per cent of the GCC region exports in recent years. There are some differences among the countries in the region in terms of the destinations of their exports, but the patterns are generally similar.

Most of the import trade is also with the industrial nations, as is clear from Table 2.15: the industrial nations collectively provided between two-thirds and three-quarters of the goods imported into the GCC region in each of the years shown in the table, and the proportion has grown in recent years. Again, there are some small differences among the individual countries.

In summary, the picture which emerges from the trade estimates presented here is that the total value of commodity exports from the GCC countries has grown very rapidly in recent years, and that this has led to rapid increases in imports, although the value of commodity imports has remained far below that of exports. While the value of trade has grown, it has continued to be mostly with the industrialized nations, which are the destination for about 70 per cent of the exports and the origin for a somewhat larger fraction of the imports.

2.3.2 *The Flow of International Payments and the Accumulation of Foreign Assets*

As a result of continuing large export surpluses, the GCC region has, for some years, accumulated large financial surpluses. Some of these funds have left the region in the form of transfers (official foreign aid, or private transfers made mostly by foreign workers), while others have involved the accumulation of assets abroad.

The peak balance of trade surplus was realized in 1980 as is clear from Table 2.16. The GCC region posted a cumulative surplus of US$107 billion. Alternatively, the current account surplus, although substantial, is drastically lower than the balance of trade surplus. This is primarily on account of high private transfers of migrant workers, official aid and deficits on services. The only country where this is not the case is Kuwait which boasted a higher current account surplus than its trade balance surplus. This is the outcome of a long tradition of investing abroad judiciously and carefully.

The accumulation of large surpluses on the balance of payment is indicative of adjustment problems. It is clear that some surplus and accumulation of foreign reserves are warrantable given the large import bills of the region. However, it is clear that the accumulated surpluses are far beyond the customary norm of three months import values. It follows that either exports are unnecessarily high or imports are unnecessarily low. In the GCC case, there is no question about the exceptionally high volume of exports that are allowed to far outstrip the import requirements of the region. In other words, the region is exporting more of its oil wealth than is warrantable by its import requirements. Given that the demand for oil is downward sloping, and generally inelastic in the relevant range, the GCC region is getting less revenues for the high volume of exports. Add to this the fact that throughout the 1970s and early 1980s the interest rate was significantly below the rate of increase in oil prices; oil was transformed into a lower wealth value. Surely, the GCC region conducted its export policy in the interest of the world community. Ironically this sacrifice has gone unnoticed and unrequited.

2.4 Human Resources

The population of the GCC region was estimated at somewhere around 12.5 million in 1981. The population is not easily measured with accuracy in economic and geographic conditions such as those that obtain within much of the region, and the quality of the population data is, therefore, variable and uncertain. Nevertheless, the 12.5 million figure at least indicates an order of magnitude.

Saudi Arabia has by far the largest share of the regional total — about 69 per cent in 1981, based on the estimates shown in Table 2.17. Almost 12 per cent live in Kuwait, and about 7 per cent in each of Oman and the United Arab Emirates. Bahrain has some 3 per cent and Qatar about 2 per cent. Saudi Arabia clearly dominates the region in terms of both geographic area and size of population.

The overall population of the region has grown rapidly within the past decade. The growth has been faster in some member countries than in others, but nevertheless rapid in all. The average rate was about 4 per cent per annum between 1971 and 1981, and roughly the same in both halves of the decade. The population of the region, which was about 8.5 million at the start of the decade, increased by some 4 million, or almost half of its original size. By the standards of the industrialized nations of the world, that is a very large proportionate increase. All other factors aside, the growth of the population alone indicates the magnitude by which the consumption requirements and the human resources of the region have been expanding.

The rapid growth in population has been attributable in large part to the inflows of foreign workers on which the region has relied in order to expand its labour force at a pace consistent with its expanding economy. Data relating to the foreign-born population in the region are fragmentary, and often out of date. Moreover, there is no question that the proportions are generally high. A 1974 estimate set the proportion at about 12 per cent in Saudi Arabia, but undoubtedly it is much higher than that today, although probably still below the levels in other GCC member countries. Figures for Kuwait and Bahrain in 1981 indicate proportions of 60 and 32 per cent, respectively. In the United Arab Emirates the proportion is still greater; even in 1975 it was estimated to have been close to 70 per cent, and all indications are that it has risen markedly since then. The implications of such large foreign-born components within the population are profound from a social point of view, as the countries of the region well recognize.

The territory of the GCC region is large and much of it is sparsely populated. Overall, the density of population is estimated to have been five persons per square kilometre in 1981. An estimate of the rural and urban proportions sets them at 67 and 33 per cent, respectively. This is based on 1976 data, and almost certainly the urban proportion has risen in the past several years. Nevertheless, a large fraction of the regional population lives in rural areas at the present time. This is due to the low urban proportion in Saudi Arabia (estimated at about 21 per cent in 1976, although one suspects that this figure may be an underestimate), and to a lesser degree to the much lower proportion in Oman (only about 5 per cent). The other four member countries have quite high urban proportions, as indicated in Table 2.17.

The population of the GCC region is a relatively young one. Again the data are of variable quality and recency, but the picture is clear in broad outline. Data for Saudi Arabia in 1974 indicate that some 47 per cent of the population were under 15 years of age, and only 11 per cent were 50 or over. One may wonder whether the under-15 proportion is overstated somewhat, especially as an indicator of the present-day situation. Even so, it is apparent that the age distribution is heavily tipped towards the younger end. In Bahrain, Kuwait and Qatar both the under-15 and 50-and-over proportions are lower, but the generalization still holds; the proportion under 15 ranges from about 33 to about 40 per cent in those three countries, based on the figures recorded in

Table 2.17, and the proportion 50 and over ranges from 5 to a little less than 9 per cent. In the United Arab Emirates the proportion under 15 was only about 28 per cent, as of 1975, and the proportion 50 and over was some 7 per cent. Thus almost two-thirds of the UAE population were in the 'prime' working age range between 15 and 50. In large measure this reflects the very high proportion of foreign-worker population in the UAE.

Women account for somewhat less than half of the overall population of the GCC region, and in some of the member countries the proportion is well below half. In Qatar the proportion is 40 per cent, and in the UAE only 31 per cent, based on 1976 data. Again, the low UAE figure reflects the very large (and preponderant male) foreign-worker component.

Levels of education have obviously not been high in the GCC region, on average, if the levels common in industrialized areas of the world are used as a standard. It is estimated that in Saudi Arabia, in 1978, about 58 per cent of the population 10 years of age and over were illiterate. In both Bahrain and Kuwait the proportion was about 36 per cent, and in the UAE it was 31. (Figures for Oman and Qatar were not available.) The high average rate of illiteracy in Saudi Arabia is a reflection, in part, of the large rural component of its population. There have been major investments in their educational sectors by the GCC countries in recent years, and substantial improvements in educational participation rates. The level of education has, therefore, been rising, on average, and the rate of illiteracy has been falling. The rates presented in Table 2.17 thus are poor indicators of the current situation with regard to the younger cohorts within the population, and hence with regard to new labour force entrants. Nevertheless, they do reflect an important characteristic of the existing population as a whole and, more importantly (from the point of view of technologically-demanding industrialization processes), of the present-day working population.

Data relating directly to the labour force are presented in Table 2.18. The quality of these data is also somewhat uncertain, and caution is warranted in interpreting them. More attention should be paid, therefore, to the broad picture of the availability and characteristics of manpower indicated by the table than to its details.

The total labour force of the region was estimated at 4.1 million in 1980, or about 33 per cent of the total population. Saudi Arabia, of course, contributed by far the largest proportion of the total — about 62 per cent, which is somewhat less than its share of the population. About 10 or 11 per cent were contributed by both Kuwait and the UAE, and another 9 per cent by Oman. Bahrain and Qatar, the least populous of the six member countries, contributed about 5 and 3 per cent respectively.

The extent of reliance on foreign workers shows up clearly in Table 2.18. In 1975, about 85 per cent of the labour force of the UAE were foreign born, and the proportion in Qatar was about 81 per cent in the same year. Bahrain had the smallest foreign proportion shown in the table — about 38 per cent (in 1976). Kuwait had about 70 per cent (in 1976), Saudi Arabia about 43 per cent (in 1975) and Oman an estimated 50 per cent (in 1980). These are very high proportions indeed, and the more so when one considers that most of the estimates are several years out of date, and that, if anything, the current proportions are likely to be higher. The desire of the countries of the GCC region to expand their economies rapidly and vigorously, created a demand for labour which could be satisfied only by opening the gates wide to migrant workers. There are probably very few areas of the world where one would find work forces with such high foreign content.

The labour force of the region, like the population, is comparatively young. Age-distribution data are not available for all countries, and once again such data as are available are out of date. But the general features are clear. In Saudi Arabia (with some three-fifths of the total GCC labour force), about 60 per cent of all workers were under the age of 35, in 1977. In the UAE the proportion was even higher — 68 per cent, in 1975. In Bahrain and Kuwait (the only other countries for which age data are shown in Table 2.18), the proportions were 54 and 56 per cent, respectively.

A striking difference between the GCC labour force and the labour forces of the industrialized nations of the world, generally, is the extremely low rates of female participation. Of the five countries for which data are shown in Table 2.18, Kuwait had the largest proportion of women in its labour force, and that was only about 12 per cent in 1975. Bahrain had a 9 per cent proportion in 1979, and Saudi Arabia a 6 per cent proportion. In Qatar and the UAE the proportions were even lower — roughly 3 per cent in each country. The reasons, of course, are cultural and, therefore, quite understandable. However, in considering strategies for economic development, and in making comparisons with industrialized nations, it is important to note that the GCC region has this particular constraint on its ability to expand its industrial productive capacity through the recruitment of workers from within the domestic population.

2.5 The Natural Resources of the Region

A brief statement of the more prominent natural resource and related environmental features of the GCC region is provided in Table 2.19. The table includes a summary description of energy resources, water resources, agriculture, fisheries and metallic and non-metallic minerals, and takes note of some other basic natural features.

The most prominent class of resources, obviously, is energy resources. These include oil and natural gas in great abundance. Indeed, the GCC region represents the main sources of crude oil exports for the rest of the world. Its vast natural gas reserves have scarcely been exploited to date, but offer great potential for the future. A major advantage of the region is that its geology makes the processes of extracting oil and gas relatively cheap. Low costs of transportation give the oil industry a further natural advantage in world markets, although this advantage is not one that is shared with gas. In addition to oil and gas as sources of cheap energy for industrial and household consumption, the region has a vast potential for the future harnessing of solar energy, owing to its climatic characteristics.

If energy resources provide the potential for industrialization, water resources represent one of the most important 'bottlenecks' — one of the chief factors operating to impede the realization of that potential. Some production processes are more water-intensive than others, of course, but in general both industrialization and agricultural expansion imply a rising level of water utilization, and the scarcity of cheap water may become an increasingly severe constraint on economic development.

Estimates for the region as a whole indicate a surplus of water supply over demand for the next few years, but the aggregates mask the existence of scarcities in many parts of the region. The largest fraction of total supply is in certain areas of Saudi Arabia; elsewhere the region is characterized generally by potential shortages. Moreover, water is largely a shared resource in the most fundamental sense: the hydrogeology of the region is such that attempts to increase the supply in one area or country may damage or reduce the supply in another. Water is critical to the economy of the region and the management and development of water resources are basic concerns for the GCC if it is to have a co-ordinated and effective policy of economic development.

The potential for developing a strong agricultural base in the GCC region is restricted by the limited availability of arable land. Most of the arable land that is available is located in Saudi Arabia and Oman, and even here the productivity of the land is heavily dependent on the controlled supply of water. Any expansion would necessitate more extensive use of irrigation, and would require careful planning and management, inasmuch as improper irrigation techniques run the risk of soil degradation and permanent loss of land from production. On the other hand, there is considerable potential for efficient production by the use of controlled-environment techniques in the growing of fruit and vegetables and in the raising of livestock, especially poultry.

The stocks of fish in the GCC region are largely concentrated in the Arabian Sea, and it is here that there exists the greatest potential for further development of fisheries. The stocks in the Arabian Gulf and the Gulf of Oman are not as large, but are nevertheless substantial, and they too offer some development potential. Stocks in the Red Sea are relatively quite small, and of comparatively minor importance from the point of view of the region as a whole.

The metallogenesis of the GCC region is associated mainly with crystalline basement outcrops. By far the largest share of known deposits of metallic minerals is located in Saudi Arabia, although there are deposits also in Oman, and to a lesser extent in the United Arab Emirates. Aside from copper in Oman, there appears to be no significant exploitation of the region's metallic mineral deposits at the present time. Metals for which extraction is possible include copper, iron, zinc, aluminium, gold and silver.

Non-metallic mineral deposits are distributed more generally throughout the GCC region. Present production from these deposits includes gypsum, limestone, clays, silica sand, gravel, and various types of ornamental stone. There are significant known deposits of phosphate rock in Saudi Arabia which have not been exploited to date, and which offer potential for future production.

The forest resources of the region are, of course, negligible. We note this in Table 2.19 simply as a reminder that the region is necessarily wholly dependent on imports for all wood materials and finished products.

A broad definition of 'natural resources' would include the locational advantage of the region associated with its access to ocean transportation. The location of the GCC countries on major maritime trade routes is an important feature of the regional economy from the point of view of its present and prospective trading relations with the rest of the world.

Two other general features of the regional environment that are of relevance in a stocktaking of natural advantages and disadvantages are noted in Table 2.19. The first has to do with the capacity of the Arabian Gulf to absorb industrial and urban waste. An unavoidable concomitant of industrialization and urbanization is an increased volume of waste that must be discharged into the environment in one form or another. The Arabian Gulf is the natural 'sink' for such discharge. However, the Gulf is shallow, highly saline and has high water temperature and slow circulation, all of which makes its capacity for absorbing waste relatively low. The problem of waste disposal is a major one for industrial nations around the world, and clearly it will be increasingly important for those of the GCC region as it moves along its path of economic development.

The second environmental feature of note is that lack of vegetation makes the soil subject to erosion in much of the GCC region, with implications for the utilization and preservation of land that introduce problems not encountered in many other parts of the world in which industrial economies have developed.

2.6 Conclusion

The picture that emerges from this synopsis is of a regional economy with substantial potential but also one that faces effective and binding constraints of skill shortages, water and a variety of natural resources. The domestic market is fragmented but purchasing power is exceptionally high; foreign workers relax skill shortages but impose high social costs in terms of congestion, pressure on service facilities and political stability. The trade-offs are clear and sharp; but so are the rewards to careful management and economic foresight.

Table 2.1: Gross Domestic Product of the GCC Region, by Sector of Origin and Country, 1971 (in current US$ million)

	Bahrain	Kuwait	Oman	Qatar	Saudi Arabia	UAE	GCC Region
Agriculture, forestry and fishing	3	9	40	7	235	20	314
Mining, quarrying and oil extraction	188	2,549	178	254	2,815	718	6,702
Manufacturing	6	152	1	11	436	31	637
Electricity, gas and water	4	26	1	5	66	15	117
Construction	5	94	49	25	224	72	469
Total, commodity-producing sectors	206	2,830	269	303	3,778	856	8,242
Trade, restaurants and hotels	14	253	7	30	239	86	629
Transportation, storage and communication	3	89	5	30	330	83	540
Other services	39	708	21	37	685	105	1,595
Total, service-producing sectors	56	1,050	33	97	1,253	274	2,763
Total GDP at producers' prices	262	3,880	301	400	5,031	1,130	11,004
Import duties					78		78
Total GDP at purchasers' prices	262	3,880	301	400	5,109	1,130	11,082
GDP *per capita* (at producers' prices)	1,189	4,912	449	3,333	801	4,036	1,310

Note: Figures in this table may not add to totals or subtotals because of rounding.
Source: The Arab Monetary Fund, *National Accounts for Arab States*, various issues.

Table 2.2: Gross Domestic Product of the GCC Region, by Sector of Origin and Country, 1976 (in current US$ million)

	Bahrain	Kuwait	Oman	Qatar	Saudi Arabia	UAE	GCC Region
Agriculture, forestry and fishing	27	35	62	23	449	109	705
Mining, quarrying and oil extraction	458	8,633	1,536	1,717	31,189	8,198	51,731
Manufacturing	142	786	12	31	2,315	150	3,436
Electricity, gas and water	10	64	14	17	43	79	227
Construction	138	419	240	316	4,491	1,508	7,112
Total, commodity-producing sectors	776	9,937	1,864	2,104	38,487	10,045	63,213
Trade, restaurants and hotels	105	988	146	216	1,751	1,031	4,237
Transportation, storage and communication	86	248	74	105	1,155	501	2,169
Other services	284	1,959	311	304	5,470	1,452	9,780
Total, service-producing sectors	475	3,195	530	625	8,376	2,984	16,185
Total GDP at producers' prices	1,251	13,132	2,394	2,729	46,863	13,030	79,399
Import duties				−25	−255	−120	−400
Total GDP at purchasers' prices	1,251	13,132	2,394	2,704	46,608	12,910	78,999
GDP *per capita* (at producers' prices)	6,015	12,389	3,031	15,021	6,298	20,822	7,655

Note: Figures in this table may not add to totals or subtotals because of rounding.
Source: The Arab Monetary Fund, *National Accounts for Arab States*, various issues.

Table 2.3: Gross Domestic Product of the GCC Region, by Sector of Origin and Country, 1981 (in current US$ million)

	Bahrain	Kuwait	Oman	Qatar	Saudi Arabia	UAE	GCC Region
Agriculture, forestry and fishing	85	62	134	47	1,648	235	2,211
Mining, quarrying and oil extraction	1,350	17,030	4,549	3,996	86,191	20,075	133,191
Manufacturing	391	1,531	64	237	5,937	1,189	9,349
Electricity, gas and water	30	98	45	73	96	368	710
Construction	544	675	361	555	15,175	2,790	20,100
Total, commodity-producing sectors	2,401	19,396	5,153	4,908	109,047	24,656	165,561
Trade, restaurants and hotels	358	1,420	365	514	6,220	2,580	11,457
Transportation, storage and communication	281	416	188	211	5,323	1,058	7,477
Other services	1,074	3,979	870	662	16,963	3,328	26,876
Total, service-producing sectors	1,713	5,815	1,423	1,387	28,506	6,965	45,809
Total GDP at producers' prices	4,113	25,212	6,577	6,294	137,554	31,621	211,371
Import duties				−92	−509	−464	−1,065
Total GDP at purchasers' prices	4,113	25,212	6,577	6,202	137,045	31,157	210,306
GDP *per capita* (at producers' prices)	10,283	17,508	7,149	26,965	15,843	36,655	16,838

Note: Figures in this table may not add to totals or subtotals because of rounding.
Source: The Arab Monetary Fund, *National Accounts for Arab States*, various issues.

Table 2.4: Index Numbers of Gross Domestic Product of the GCC Region at Constant Prices, by Sector of Origin (base year 1970 = 100)

Year	Agriculture, forestry and fishing	Mining, quarrying and oil extraction	Manufacturing	Electricity, gas and water	Construction	Wholesale and retail trade	Transport and communication	Other	Total GDP
1960	69	40	36	25	44	44	30	53	43
1963	88	51	46	50	59	54	44	67	55
1965	92	63	54	64	71	65	56	81	67
1970	100	100	100	100	100	100	100	100	100
1971	103	117	113	110	116	110	113	125	117
1972	106	135	120	119	104	121	133	132	129
1973	109	153	135	130	115	140	183	150	145
1974	116	154	169	143	150	178	225	166	159
1975	125	146	169	138	211	220	225	183	167
1976	133	154	191	181	252	286	288	219	192
1977	138	158	214	219	324	349	353	240	212
1978	153	161	232	297	376	345	408	259	225
1979	157	189	248	361	383	371	432	284	249
Annual Average Percentage Growth Rates									
1960-65	5.9	9.5	8.4	20.7	10.0	8.1	13.3	8.9	9.3
1965-70	1.7	9.7	13.1	9.3	7.1	9.0	12.3	4.3	8.3
1970-75	4.6	7.9	11.1	6.7	16.1	17.1	17.6	12.8	10.8
1975-79	5.9	6.7	10.1	27.2	16.1	14.0	17.7	11.6	10.5
1960-79	4.4	8.5	10.7	15.1	12.1	11.9	15.1	9.2	9.7
1970-79	5.1	7.3	10.6	15.3	16.1	15.7	17.7	12.3	10.7

Source: United Nations, Economic Commission for Western Asia, *National Accounts Studies Bulletin, No. 4* (October 1981).

Table 2.5: Expenditures on Gross Domestic Product, 1971 (in current US$ million)

	Bahrain	Kuwait	Oman	Qatar	Saudi Arabia	UAE	GCC Region
Private consumption	230	1,179	51	70	1,429	153	3,112
Public consumption	59	487	63	69	847	150	1,675
Total consumption	289	1,666	114	139	2,276	303	4,787
Investment	18	364	86	59	608	306	1,441
Exports of goods and services (X)	285	2,574	198	311	3,385	957	7,710
Imports of goods and services (M)	330	724	97	109	1,160	437	2,857
Balance of trade (X − M)	−45	1,850	101	202	2,225	520	4,853
Total GDP (at purchasers' prices)	262	3,880	301	400	5,109	1,130	11,082

Note: Figures in this table may not add to totals or subtotals because of rounding.
Source: The Arab Monetary Fund, *National Accounts for Arab States*, various issues.

Table 2.6: Expenditures on Gross Domestic Product, 1976 (in current US$ million)

	Bahrain	Kuwait	Oman	Qatar	Saudi Arabia	UAE	GCC Region
Private consumption	842	3,522	304	176	6,771	1,947	13,598
Public consumption	243	1,479	705	452	8,182	1,176	12,237
Total consumption	1,085	5,001	1,045	628	14,953	3,122	25,834
Investment	290	2,172	930	699	9,722	4,284	18,097
Exports of goods and services (X)	1,629	10,234	1,596	2,209	34,075	9,134	58,877
Imports of goods and services (M)	1,753	4,275	1,177	833	12,143	3,630	23,811
Balance of trade (X − M)	−124	5,959	419	1,376	21,932	5,503	35,065
Total GDP (at purchasers' prices)	1,252	13,132	2,394	2,704	46,608	12,910	79,000

Note: Figures in this table may not add to totals or subtotals because of rounding.
Source: The Arab Monetary Fund, *National Accounts for Arab States*, various issues.

Table 2.7: Expenditures on Gross Domestic Product, 1981 (in current US$ million)

	Bahrain	Kuwait	Oman	Qatar	Saudi Arabia	UAE	GCC Region
Private consumption	2,920	8,374	942	281	23,935	6,429	42,881
Public consumption	667	3,641	1,332	623	31,989	3,259	41,511
Total consumption	3,587	12,015	2,274	904	55,924	9,688	84,392
Investment	553	4,007	1,621	1,078	35,883	8,296	51,438
Exports of goods and services (X)	4,951	19,472	5,032	5,625	93,522	23,259	151,861
Imports of goods and services (M)	4,977	10,283	2,351	1,406	48,284	10,085	77,386
Balance of trade (X − M)	−26	9,189	2,681	4,219	45,238	13,174	74,475
Total GDP (at purchasers' prices)	4,113	25,211	6,577	6,202	137,045	31,157	210,305

Note: Figures in this table may not add to totals or subtotals because of rounding.
Source: The Arab Monetary Fund, *National Accounts for Arab States*, various issues.

Table 2.8: Gross Domestic Product of the GCC Region, by Sector of Origin and Country: Percentage Distributions Among Countries, 1971

	Bahrain	Kuwait	Oman	Qatar	Saudi Arabia	UAE	Total
Agriculture, forestry and fishing	1.0	2.9	12.7	2.2	74.8	6.4	100.0
Mining, quarrying and oil extraction	2.8	38.0	2.7	3.8	42.0	10.7	100.0
Manufacturing	0.9	23.9	0.2	1.7	68.4	4.9	100.0
Electricity, gas and water	3.4	22.2	0.9	4.3	56.4	12.8	100.0
Construction	1.1	20.0	10.4	5.3	47.8	15.4	100.0
Total, commodity-producing sectors	2.5	34.3	3.3	3.7	45.8	10.4	100.0
Trade, restaurants and hotels	2.2	40.2	1.1	4.8	38.0	13.7	100.0
Transportation, storage and communication	0.6	16.5	0.9	5.6	61.1	15.4	100.0
Other services	2.4	44.4	1.3	2.3	43.0	6.6	100.0
Total, service-producing sectors	2.0	38.0	1.2	3.5	45.4	9.9	100.0
Total GDP at producers' prices	2.4	35.3	2.7	3.6	45.7	10.3	100.0
Import duties					100.0		
Total GDP at purchasers' prices	2.4	35.0	2.7	3.6	46.1	10.2	100.0

Note: Figures in this table may not add to totals or subtotals because of rounding.
Source: Table 2.1, p. 23.

Table 2.9: Gross Domestic Product of the GCC Region, by Sector of Origin and Country: Percentage Distributions Among Countries, 1976

	Bahrain	Kuwait	Oman	Qatar	Saudi Arabia	UAE	Total
Agriculture, forestry and fishing	3.8	5.0	8.8	3.3	63.7	15.5	100.0
Mining, quarrying and oil extraction	0.9	16.7	3.0	3.3	60.3	15.8	100.0
Manufacturing	4.1	22.9	0.3	0.9	67.4	4.4	100.0
Electricity, gas and water	4.4	28.2	6.2	7.5	18.9	34.8	100.0
Construction	1.9	5.9	3.4	4.4	63.2	21.2	100.0
Total, commodity-producing sectors	1.2	15.7	3.0	3.3	60.9	15.9	100.0
Trade, restaurants and hotels	2.5	23.3	3.5	5.1	41.3	24.3	100.0
Transportation, storage and communication	4.0	11.4	3.4	4.8	53.3	23.1	100.0
Other services	2.9	20.0	3.2	3.1	55.9	14.9	100.0
Total, service-producing sectors	2.9	19.7	3.3	3.9	51.8	18.4	100.0
Total GDP at producers' prices	1.6	16.5	3.0	3.4	59.0	16.4	100.0
Import duties				6.2	63.8	30.0	100.0
Total GDP at purchasers' prices	1.6	16.6	3.0	3.4	59.0	16.3	100.0

Note: Figures in this table may not add to totals or subtotals because of rounding.
Source: Table 2.2, p. 24.

Table 2.10: Gross Domestic Product of the GCC Region, by Sector of Origin and Country: Percentage Distributions Among Countries, 1981

	Bahrain	Kuwait	Oman	Qatar	Saudi Arabia	UAE	Total
Agriculture, forestry and fishing	3.8	2.8	6.1	2.1	74.5	10.6	100.0
Mining, quarrying and oil extraction	1.0	12.8	3.4	3.0	64.7	15.1	100.0
Manufacturing	4.2	16.4	0.7	2.5	63.5	12.7	100.0
Electricity, gas and water	4.2	13.8	6.3	10.3	13.5	51.8	100.0
Construction	2.7	3.4	1.8	2.8	75.5	13.9	100.0
Total, commodity-producing sectors	1.4	11.7	3.1	3.0	65.9	14.9	100.0
Trade, restaurants and hotels	3.1	12.4	3.2	4.5	54.3	22.5	100.0
Transportation, storage and communication	3.8	5.6	2.5	2.8	71.2	14.1	100.0
Other services	4.0	14.8	3.2	2.5	63.1	12.4	100.0
Total, service-producing sectors	3.7	12.7	3.1	3.0	62.2	15.2	100.0
Total GDP at producers' prices	1.9	11.9	3.1	3.0	65.1	15.0	100.0
Import duties				−8.6	−47.8	−43.6	−100.0
Total GDP at purchasers' prices	2.0	12.0	3.1	2.9	65.2	14.8	100.0

Note: Figures in this table may not add to totals or subtotals because of rounding.
Source: Table 2.3, p. 25.

Table 2.11: Commodity Trade Within the GCC Region and With the Rest of the World, 1981 (in current US$ million)

From/To	Bahrain	Kuwait	Oman	Qatar	Saudi Arabia	UAE	Other GCC	Rest of World	Total exports
Bahrain	–	32.2	78.6	33.4	146.3	635.5	926.1	2,615.1	3,541.2
Kuwait	5.0	–	4.0	25.0	132.0	411.0	577.0	15,984.0	16,561.0
Oman	...	...	...	...	...	...	...	4,415.8	4,415.8
Qatar	2.6	19.6	1.1	–	69.4	45.9	138.8	3,839.3	3,978.1
Saudi Arabia	2,544.0	191.0	19.0	23.0	–	78.0	2,855.0	110,473.0	113,328.0
UAE	20.0	12.0	272.0	38.0	86.0	–	428.0	20,511.0	20,939.0
Other GCC	2,571.6	255.1	374.7	119.4	433.7	1,170.4	–	–	–
Rest of World	1,814.1	7,786.9	1,846.1	1,451.4	34,834.3	8,378.6	–	–	–
Total imports	4,385.7	8,042.0	2,220.8	1,570.8	35,268.0	9,549.0	–	–	–

Note: Where possible, information is taken from export data. The symbol – indicates 'not relevant'; ... indicates 'not available'.
Source: International Monetary Fund, *Direction of Trade Statistics Yearbook*, 1982.

Table 2.12: Exports from the GCC Region, Total and Percentage Distribution Among Countries, 1975–1981

Year	Total (US$ million)	Percentage Distribution						
		Bahrain	Kuwait	Oman	Qatar	Saudi Arabia	UAE	GCC Region
1975	47,563.7	1.1	18.1	2.8	3.7	62.0	12.2	100.0
1976	60,618.6	1.2	16.2	2.5	3.5	62.8	13.8	100.0
1977	66,767.6	1.3	14.6	2.3	3.3	64.5	14.0	100.0
1978	60,859.3	1.2	17.1	2.4	3.7	61.3	14.4	100.0
1979	95,771.0	1.3	19.1	2.2	3.4	60.5	13.6	100.0
1980	151,062.1	1.0	13.4	1.9	2.9	67.1	13.6	100.0
1981	157,838.2	0.6	10.3	2.6	2.4	71.5	12.5	100.0

Source: International Monetary Fund, *Direction of Trade Statistics Yearbook*, 1982.

Table 2.13: Imports to the GCC Region, Total and Percentage Distribution Among Countries, 1975–1981

Year	Total (US$ million)	Percentage Distribution						
		Bahrain	Kuwait	Oman	Qatar	Saudi Arabia	UAE	GCC Region
1975	10,403.9	5.3	22.9	5.1	3.6	37.7	25.4	100.0
1976	16,405.3	5.7	20.2	3.2	4.6	46.6	19.8	100.0
1977	26,365.9	4.1	18.3	2.7	4.3	51.8	18.7	100.0
1978	32,983.0	3.4	13.9	2.3	3.5	61.1	15.8	100.0
1979	39,051.5	3.1	13.1	2.6	3.5	61.3	16.4	100.0
1980	48,656.9	3.0	13.2	2.8	2.8	61.3	16.9	100.0
1981	56,262.6	2.8	13.8	3.2	2.6	61.8	15.7	100.0

Source: International Monetary Fund, *Direction of Trade Statistics Yearbook*, 1982.

Table 2.14: Exports from the GCC Region to Industrial and Developing Countries, Percentage Shares, 1975–1981

		Bahrain	Kuwait	Oman	Qatar	Saudi Arabia	UAE	GCC Region
1975	— industrial	51.7	62.4	83.2	72.1	68.3	92.7	70.4
	— developing	22.4	30.2	16.6	26.1	22.0	4.9	21.3
	— non-GCC Arab countries	0.0	4.7	...	0.1	1.2	0.0	1.6
1976	— industrial	35.9	57.7	90.9	75.6	66.5	85.3	67.9
	— developing	25.0	28.3	8.9	16.1	22.8	9.3	21.3
	— non-GCC Arab countries	0.9	22.2	...	1.2	1.3	0.1	1.3
1977	— industrial	30.6	58.8	89.2	69.7	69.5	81.3	69.0
	— developing	26.3	25.2	10.1	14.4	21.6	13.9	20.7
	— non-GCC Arab countries	1.6	2.2	...	2.3	1.6	0.4	1.5
1978	— industrial	35.8	62.2	93.0	77.4	75.2	75.8	72.6
	— developing	32.3	23.0	6.0	21.6	18.6	18.6	19.5
	— non-GCC Arab countries	1.7	2.6	0.0	...	1.3	1.7	1.5
1979	— industrial	21.3	60.0	94.9	73.2	75.7	76.8	71.9
	— developing	27.8	27.3	1.2	22.2	19.1	20.5	20.8
	— non-GCC Arab countries	1.1	3.5	...	0.0	1.8	2.1	2.1
1980	— industrial	23.0	50.3	77.6	70.5	75.3	81.0	71.5
	— developing	30.1	33.9	15.7	25.7	19.8	16.6	21.5
	— non-GCC Arab countries	1.5	4.0	...	0.0	1.9	1.6	2.0
1981	— industrial	22.5	43.6	83.0	66.7	72.4	77.2	69.1
	— developing	30.5	38.0	9.8	28.7	22.6	20.8	23.9
	— non-GCC Arab countries	1.2	4.8	...	0.0	2.5	1.7	2.5

Note: The symbol ... indicates 'not available'.
Source: International Monetary Fund, *Direction of Trade Statistics Yearbook*, 1982.

Table 2.15: Imports to the GCC Region from Industrial and Developing Countries, Percentage Shares, 1975-1981

		Bahrain	Kuwait	Oman	Qatar	Saudi Arabia	UAE	GCC Region
1975	— industrial	35.2	77.1	64.9	76.9	63.7	69.0	65.3
	— developing	9.6	18.4	14.6	14.4	25.3	20.0	20.0
	— non-GCC Arab countries	0.5	3.8	1.1	5.6	18.0	2.1	8.1
1976	— industrial	41.0	74.9	62.3	78.4	64.3	73.9	66.5
	— developing	13.4	20.8	15.2	10.2	20.0	18.9	18.8
	— non-GCC Arab countries	0.3	2.3	0.4	2.7	12.4	1.3	6.6
1977	— industrial	39.7	71.8	66.5	81.1	66.4	75.2	67.6
	— developing	12.3	23.8	15.0	10.5	16.3	17.6	17.2
	— non-GCC Arab countries	0.3	2.2	0.4	2.3	7.6	1.2	4.6
1978	— industrial	42.3	74.1	63.9	85.5	79.9	75.8	76.0
	— developing	11.4	20.7	16.5	11.1	12.8	17.0	14.5
	— non-GCC Arab countries	0.6	3.0	0.4	2.5	3.5	1.4	2.8
1979	— industrial	38.3	72.5	67.5	84.6	79.1	71.9	74.5
	— developing	8.7	21.7	13.7	11.5	13.8	17.3	15.0
	— non-GCC Arab countries	0.3	2.6	0.5	2.2	3.4	1.1	2.6
1980	— industrial	32.2	74.1	63.4	77.7	79.6	70.7	73.7
	— developing	8.7	20.0	12.8	15.4	14.5	19.6	15.7
	— non-GCC Arab countries	0.2	2.3	0.3	2.5	3.2	3.1	2.8
1981	— industrial	27.7	74.7	67.8	78.9	80.6	80.9	74.0
	— developing	7.2	18.8	12.2	14.1	14.3	19.7	15.1
	— non-GCC Arab countries	0.1	1.8	0.2	1.8	3.1	2.8	2.6

Source: International Monetary Fund, *Direction of Trade Statistics Yearbook*, 1982.

Table 2.16: Balance of Payments of the GCC Member Countries, 1979–1981 (in current US$ million)

	Balance of Trade			Current Account			Overall Balance		
	1979	1980	1981	1979	1980	1981	1979	1980	1981
Bahrain	−47	266	139	−66	389	211	63	531	809
Kuwait	13,242	13,835	8,629	14,204	15,799	13,760	11,093	12,418	489
Oman	783	1,683	2,120	439	1,248	1,181	323	1,225	1,149
Qatar	2,459	4,244	4,357	1,609	2,710	2,492	851	1,569	839
Saudi Arabia	32,991	72,478	76,700	9,589	39,799	45,100	130	4,862	10,987
UAE	8,963	14,728	13,157	5,529	14,066	8,989	2,411	4,856	2,362
GCC total	58,391	107,234	105,102	31,304	74,011	71,733	14,871	25,461	16,635

Source: Arab Fund for Economic and Social Development, Arab League, Arab Monetary Fund and OAPEC, *The Joint Arab Economic Report*, 1982, p. 205; 1983, p. 265.

Table 2.17: Size and Selected Characteristics of the Population of the GCC Region, by Country, Various Years

	Bahrain	Kuwait	Oman	Qatar	Saudi Arabia	UAE	GCC Region
Total population ('000)							
– 1971	220	790	670	120	6,380	280	8,460
– 1976	270	1,060	790	180	7,400	620	10,320
– 1981	400	1,440	920	230	8,650	850	12,490
Annual growth rate (%)							
– 1971-81	6.2	6.2	3.2	6.7	3.1	11.7	4.0
– 1971-76	4.2	6.1	3.3	8.4	3.0	17.2	4.1
– 1976-81	8.2	6.3	3.1	5.0	3.2	6.5	3.9
Country as % of region, 1981	3.2	11.5	7.4	1.8	69.3	6.8	100.0
Density (pop./km²), 1981	597.6	80.8	3.9	20.2	4.0	10.3	5.0
Urban as % of total, 1976	77.7	88.4	5.3	88.7	20.8	84.0	33.0
Female as % of total, 1976	47.2	46.0	49.2	40.0	49.4	30.8	47.7
Foreign as % of total (latest year)	32.0 (1981)	60.0 (1981)	n.a.	61.8 (1975)	11.8 (1974)	69.5 (1975)	n.a.
% Illiterate, pop. 10+, 1978	36.1 (1981)	36.5 (1975)	n.a.	n.a. (1970)	57.9 (1974)	31.0 (1975)	n.a.
% Age distribution (latest year)							
– under 15	32.9	39.6	n.a.	36.8	46.7	28.2	n.a.
– 15-49	58.4	55.4	n.a.	56.4	42.3	64.5	n.a.
– 50+	8.7	5.0	n.a.	6.9	11.1	7.3	n.a.

Source: (1) United Nations, Economic Commission for Western Asia, *Statistical Indicators of the Arab World for the Period 1970-1978*.
(2) United Nations, Economic Commission for Western Asia, *The Population Situation in the ECWA Region*, various countries and various issues.
(3) The Arab Fund for Economic and Social Development, *Annual Arab Report, 1981*.

Table 2.18: Size and Selected Characteristics of the Labour Force of the GCC Region, by Country, 1981

	Bahrain	Kuwait	Oman	Qatar	Saudi Arabia	UAE	GCC Region
Labour force ('000), 1981	190	430	370	110	2,550	450	4,100
Country as % of region, 1981	4.6	10.5	9.0	2.7	62.2	11.0	100.0
Foreign as % of total	37.5	69.7	50.0	81.1	43.0	84.8	n.a.
(latest year)	(1976)	(1976)	(1980)	(1975)	(1975)	(1975)	
% Age distribution (latest year)	(1971)	(1975)			(1977)	(1975)	
– under 25	24.6	21.4	n.a.	n.a.	24.9	26.4	n.a.
– 25–34	29.4	35.0	n.a.	n.a.	34.9	41.2	n.a.
– 35–49	30.7	34.1	n.a.	n.a.	28.2	25.7	n.a.
– 50+	15.3	9.4	n.a.	n.a.	12.0	6.8	n.a.
Females as % of total	9.3	11.6		2.9	6.0	3.4	n.a.
(latest year)	(1979)	(1975)		(1970)	(1980)	(1975)	

Source: (1) United Nations, Economic Commission for Western Asia, *Statistical Abstract of the Region of the Economic Commission for Western Asia, 1970–1979.*
(2) Saudi Arabia, Ministry of Planning, *Third Development Plan, 1980–1985.*
(3) Oman, Development Council, *The Second Five-Year Plan, 1981–1985.*
(4) Arab Planning Institute and the International Labour Organization, *Seminar on Population, Employment and Migration in the Arab Gulf States,* Kuwait, December 1978.

Table 2.19: Resource Base and Related Environmental Features of the GCC Region: An Overview

Energy resources	— abundant oil and natural gas reserves — main source of world exports of crude oil — natural gas reserves so far little exploited, great potential — low costs of production for oil and gas for geological reasons — low transport cost in the case of oil — excellent potential for development of solar energy — energy resources provide main source of financial capital for future industrialization
Water resources	— GCC region characterized by relative scarcity of water supply — largest part of water supply is in Saudi Arabia; imbalances of supply and demand throughout the GCC region — water is largely a shared resource — critical element for further economic development
Agriculture	— very limited availability of arable land; most of arable land is in Saudi Arabia and Oman and is heavily dependent on irrigation — irrigation is vital but soil degradation may result from use of improper procedures and scarcity of water may be a limiting factor — high potential for controlled-environment production of fruit and vegetables — high potential for controlled-environment production of livestock, especially poultry
Fisheries	— largest fish stocks and potential for increased utilization are in the Arabian Sea; stocks and potential in the Arabian Gulf and the Gulf of Oman are smaller but substantial; stocks in the Red Sea are relatively quite small
Metallic minerals	— metallogenesis in the GCC region is associated mainly with crystalline basement outcrops — Saudi Arabia has largest share of known deposits; some known deposits also in Oman, and to a lesser extent in UAE — principal metals that could be extracted from known deposits include copper, iron, zinc, aluminium, gold and silver — significant exploitation is confined to copper deposits in Oman

Table 2.19: Resource Base and Related Environmental Features of the GCC Region: An Overview *(cont.)*

Non-metallic minerals	— known deposits are distributed generally throughout the GCC region
	— minerals now produced include gypsum, limestone, clays, silica sand, gravel, and various types of ornamental stone
	— there are also significant known deposits of phosphate rock located in Saudi Arabia which might be exploited
Forest resources	— negligible
Transportation	— countries of the GCC region are located on major maritime trade routes
Other environmental features	— Arabian Gulf is shallow, highly saline, has high water temperature and slow circulation; hence absorptive capacity is relatively low for industrial and urban waste
	— lack of vegetation makes soil subject to wind erosion in much of the GCC region

3 OIL AND GAS

3.1 Introduction

The dominance of oil in the economies of the Arabian Gulf is well known. What is perhaps less known and appreciated is the importance and role of gas in the region and the many changes that are likely to emerge as these countries co-ordinate their production, pricing, exports and industrialization of oil.

Together the GCC member countries account for a formidable 42 per cent of world proven oil reserves and for about 25 per cent of world natural gas reserves. Collectively, they account for over 40 per cent of the world oil exports and are now on the threshold of entering the world petrochemical industry in a decisive manner that will bring about substantial restructuring of this industry.

3.2 Oil

Each of the GCC member states accounts for a substantial share of the world energy supply and the collective share is very large. Although the region has less than 1 per cent of the world's oil wells, it accounted for 42 per cent of the world's proven reserves of crude oil in 1980, and about 63 per cent of the corresponding OPEC total. A decade earlier, the GCC share of world proven crude oil reserves was only about 18 per cent, or some 113 billion barrels. By 1978, a peak year, these reserves had risen to an estimated 276 billion barrels (see Table 3.1, p. 53).

The crude oil production share of the GCC is not as high as its share of proven reserves, but the region is nevertheless the world's largest producer of crude oil. In 1980, the GCC daily production of crude averaged 14.1 million barrels, or about 23.5 per cent of the world total and over 52.3 per cent of OPEC's total (see Table 3.2).[3] The difference between reserve share and production share has its roots in the political decisions of the old oil consortium, but more recently reflects the conservation decisions of the member states. The allocation in the past of low production rates to the Gulf wells by the oil consortium has left the area with a longer production life span than would otherwise have been the case. In 1980, the GCC reserves were equivalent to 53 years of production at the current rate. By comparison, the average production life of the OPEC reserves was 44 years, and of total world reserves only 30 years in 1980.

Saudi Arabia is the principal oil producer in the region and commands the largest reserve pool. In 1980, the Saudi production of 9.9 million barrels per day accounted for over 70 per cent of total GCC production, and Saudi proven reserves were 62 per cent of the estimated total. The smallest producer is Bahrain and it is also the state with the smallest reserves. Most of Bahrain's crude oil production is diverted to the local refinery. However, in the GCC as a whole, most of the crude oil produced is exported, and the region thus accounts for the major part of world trade in crude oil. In 1979, the GCC region's share of world exports of crude oil was 40 per cent; in 1975 the figure had been as high as 48.5 per cent. The GCC's share of total OPEC exports of crude oil shows a similar trend, having declined from a high of 57.5 per cent in 1975 to slightly over 50 per cent in 1979.

The large volume of crude oil exports from the GCC region is the result of a number of fundamental factors. First and foremost is the low cost of production and, recently, of transporting GCC crude oil. Most oil wells in the region are large, close to the surface, largely free flowing, optimally penetrated and close to coastal outlets. Average productivity of oil wells in the region is estimated to be over 400 times the world average (centrally planned economies excluded). Consider, for instance, the highly productive Burgan field in south-east Kuwait which is only 14-20 miles from the Gulf. A network of gathering lines connects all wells in this field to the storage tanks at Al-Ahmadi on a ridge which is 400 feet above sea level, six miles from the Gulf. Pipelines from the storage farms use gravity to feed oil to the terminal at Al-Ahmadi. (Loading by gravity rather than by use of pumps results in substantial cost savings and, at the same time, in faster loading rates.) This case is typical of most oil wells in the region, including Dukhan (Qatar), Zakum (UAE), Ghawar (Saudi Arabia), and others. Additional advantage has been provided by the falling unit costs of transporting oil across the oceans as larger and larger supertankers have been brought into use. This has been particularly important with respect to the US and Japanese markets, where Arabian Gulf oil can be delivered at prices below the corresponding domestic production costs and/or competitors' costs.

Second, GCC crude oil producers are typically believed to be discretionary producers, that is, they can reduce or increase supply without any major consequences for their economies. The GCC member states, with varying degrees of freedom, could open or close the oil tap at government discretion since their production has often been below shut-in capacity and above their foreign exchange requirements. However, there have been two restraining and countervailing forces. Within the range of discretionary oil production, the actual level appears to have depended less on the expected rates of return and differences of risk on recycled assets than on the evaluation of the likely effects that discretionary change might have on the world economy and the world political scene. Furthermore, the large financial surpluses forced on these economies raised their stake in maintaining the value of these assets by avoiding the precipitation of large changes in oil production and/or prices.

Were oil resources everlasting and renewable, the citizens of the GCC would be entitled to a perpetual rent accruing from these resources and economic diversification would not be a critical consideration. But, as already mentioned, oil supplies are finite and non-renewable. At recent rates of utilization, oil in the GCC will run out in the lifetime of the present generation (Qatar, Bahrain, Oman), its children (UAE), or its grandchildren (Saudi Arabia and Kuwait). The accumulation of large financial surpluses in the early 1970s preceded any deliberate plan for their domestic absorption or investment abroad. There was no historical experience on which to base predictions of further growth in domestic absorption and, therefore, no reason to question the economic rationality of a surplus of the magnitude realized. However, GCC producers are no longer oblivious to the risks of accumulating fixed-income-yielding assets in an inflationary world. They are, therefore, no longer satisfied with their role as just the residual suppliers of the world crude oil requirements. Now they are contemplating the creation of an advanced and integrated industrial base and the expansion of their sphere of control over the transportation, refining, processing and marketing of their oil and its derivatives.

The utilization of oil resources in 'downstream' activities will be considered, following our discussion of natural gas, as oil and gas are often joint products (associated gas) or joint inputs (in petrochemical processes).

3.3 Natural Gas Resources

Underground oil reservoirs contain dissolved natural gases as a result of pressure and heat, and when oil flows to the surface these gases are freed. Associated with each barrel of oil is a certain volume of gases — about 500 cubic feet, on average, in most GCC member countries. These are known as associated gases and possess characteristics that vary according to the circumstances of the reservoir. To avoid losing these gases, crude oil production is received at separator and cooling units, where the heavier fractions are separated. The remaining lighter gases come out as gas mix, some of which is reinjected into the oil fields for reasons of production; the other part is used locally or flared into the atmosphere.

All of the crude oil producers in the GCC region, by virtue of their oil production, have a potential command over associated gas. The realization of this potential was delayed by a number of factors, most important of which were the low price of oil and the foreign control over oil production. With the rise in oil price and the assumption of control of production and profits following 1973, the proportion of associated gas that is flared declined sharply. As shown in Table 3.4, the proportion of waste in 1971 for the GCC region as a whole (Oman and Bahrain excepted) was 75.6 per cent of the region's gross gas production, and the UAE's percentage was as high as 90 per cent. By 1980, however, the waste proportion had fallen to 55.9 per cent in the region as a whole. Moreover, the high regional percentage is heavily weighted by Saudi Arabia's experience, and Saudi flaring is expected to decline sharply with the completion of the large complexes at Al-Jubail and Yanbu. This will certainly imply a drastic reduction in the percentage waste of gas produced in the GCC region. Kuwait's, Qatar's and Bahrain's percentages of wasted gas are already quite low.

With the decline in the flaring of gas, gross production figures will reflect more accurately the rates of utilization of reserves. Gross production figures are presented in Table 3.5 for all GCC member states but Oman. The GCC total increased from 59,159 million cubic metres in 1971 to 87,019 in 1980. Saudi production had more than doubled by 1980, whereas Kuwaiti production was less than half its 1971 level. GCC production of associated gas as a percentage of OPEC or world production did not vary much throughout the period 1971-1980: in 1971, the GCC production share was 30 per cent of OPEC's and in 1980 it was 32.2 per cent; similarly, the GCC's share of world total production was 5.0 per cent in 1971 and 5.6 per cent in 1980.

Another type of natural gas is produced from reservoirs with no crude oil, and its production is easier to control. GCC proven reserves of this type of gas are believed to be immense. Indeed, some analysts have put the figure as high as 25 per cent of the world total. Much of the recent finds are in Qatar, the UAE and Bahrain. As yet, though, no firm figures have been released.

Published estimates of GCC proven reserves of natural gas place the total in the neighbourhood of 5,426 billion cubic metres in 1980. This represents about 22.5 per cent of OPEC's total reserves and about 7.4 per cent of the world total. Again, although proven reserves in the region doubled between 1971 and 1980, so did the OPEC and world reserves. Thus, GCC percentage shares of proven reserves were not much different in 1980 from what they had been in 1971. The largest present reserves are in Saudi Arabia and in Qatar's North West Dome.

The separate availability of natural gas in abundance in the region adds a new dimension to its energy stature in the world context. But more importantly, it enables the region to move more solidly into petrochemical production since gas feedstocks into this industry are noted for being superior to refined petroleum.

3.4 Present Utilization and Further Processing of GCC Oil and Gas Resources

The dramatic increase in the national incomes of the GCC states as oil prices increased, still left their economies — outside the oilfields — in a relative state of underdevelopment. Levels of living in the region have certainly risen, but essentially and primarily through a form of capital consumption, namely the depletion of oil reserves.

The GCC member states are increasingly aware of the fact that any major devaluation of their oil would likely bring financial difficulties, and that at any rate their oil is a finite resource whose end is in sight of even the present generation for at least three member states — Qatar, Bahrain and Oman, as was noted earlier.

Were oil revenues to last forever, as was mentioned above, there would be no need for diversification or concern about alternative sources of income. A continuous stream of rent would accrue to the government, which it could in turn pass on to the citizens as dividends. But this unfortunately is not the case. With oil revenues variable and finite, alternative sources of income are necessary. At the same time, ways and means must be examined to upgrade the revenues from the dwindling supplies of oil and gas. The focus of this chapter is on upgrading the value-added component of the gross output of hydrocarbons. Diversification issues, although not separable from the upgrading issue, are different, and will be treated

46 *Oil and Gas*

in a later chapter.

The GCC states are no longer satisfied with their role as crude oil producers. They are moving to build a vertically integrated industrial structure covering transportation, processing, refining and marketing activities. We will concentrate here on refining and processing, but this does not mean that the other aspects are any less important; they are simply not directly related to the issues of this chapter.

3.4.1 Petroleum Refining

Crude oil by itself does not have direct applications. Its full value is realized after it is processed into refined products for specific end uses. Furthermore, refining is a necessary first step for downstream development of fuel and non-fuel uses. In a series of processes, crude oil is converted into different fuels for energy uses as well as into lubricants, asphalt, waxes, gasoil and naphtha. The last two products are basically feedstocks for the petrochemical industry.

There were less than 900 operational refineries in the world in 1980, with a combined capacity of about 80 million barrels per day. Refineries in the GCC region currently number 13, with a combined capacity of 1.5 million barrels per day, or about 1.9 per cent of the world capacity. There are a number of new refineries planned and some are already under construction. By 1986 these would raise the GCC capacity to 3.4 million barrels per day. With this increased refining capacity, the GCC countries, with an output capacity of over 15 million barrels per day of crude, could be refining almost a quarter of their production (see Table 3.9).

Refining in the GCC region started in Bahrain in 1937 with a 25,000 b/d complex, followed in 1949 by the 25,000 b/d Al-Ahmadi plant in Kuwait, and then by Saudi Arabia's Ras Tanura. These three refineries were, and still are, the largest refining centres in the Arab world. Currently, refining capacity in the region varies between 6,300 b/d (Umm Said) in Qatar to 500,000 b/d (Ras Tanura) in Saudi Arabia. The location, capacity and type of each refinery in existence or planned are specified in Table 3.10.

The combined output of the GCC refineries will be more than sufficient to meet the expected domestic demand for refined products and, therefore, will allow for exports. However, unit transport costs are much higher for refined petroleum than for crude oil. Thus, potential GCC exports will depend on transport capacity and the ability to effect reductions in crude oil exports. In 1980, the transport cost differentials between crude and refined oil per barrel were $1.44 to the US East Coast, $1.58 to Japan, $1.29 to north-west Europe and $1.01 to southern Europe.[4] These differences in transport costs translate into refining cost differentials between the GCC and consuming countries of $1.30 with the US East Coast, $1.66 with Japan, $1.30 with north-west Europe and $1.00 with southern Europe. The differentials are not expected to fall before 1985. Thus, GCC countries are likely to face some difficulty in supplying refined products, except on a supply-demand balancing basis (i.e. filling gaps). Thus, a decision to expand refining capacity should be coupled with a decision to reduce exports of crude petroleum. Increased production should be directed to markets where competitive supplies are limited and capacity to transport the products on domestic ships should be expanded, so as to counteract conference shipping rates which discriminate against processed products. Moreover, there are a number of other issues that need to be considered in the decision to expand refining capacity. In particular, there is the issue of industrial linkages in both directions, forward and backward. Complementarity with petrochemical processes needs to be considered and feedback effects on engineering and design skills should be taken into account. Capital cost differentials between the Gulf and OECD countries (now about 50 per cent) should be reduced, where possible, by increased domestic involvement in the conception, design, procurement, installation and operation of refineries. The experience of Iraq with regard to the Basra refinery suggests that a significant reduction in capital costs may be effected through reliance on domestic capabilities.

3.4.2 Natural Gas Liquefaction

The price of oil was relatively low before 1973, and as a result no capital investments were made to exploit the associated gas through liquefaction and export. After 1973, capital investments became economically feasible and many GCC countries moved to utilize their flared gases. Some started to liquefy methane and ethane, only to discover that this was costly and returned low net profits. Liquefaction at low temperatures (260°F below zero for methane, 129°F below zero for ethane), the use of refrigerated carriers, and the need to change liquid gas back to its gaseous state at the points of destination, proved too costly to net enough return on the large capital investments required. More recently, there has been a concentration on the liquefaction of propane, butane and natural gasoline.

Existing liquid natural gas production capacity in the GCC region is in the neighbourhood of 23.2 thousand tons per year. Allowing for planned expansion, capacity in 1985 is expected to exceed 42.5 thousand tons (see Table 3.11). Propane production is expected to be 15 thousand tons per year, whereas butane and natural gasoline will each account for 11.2 thousand tons per year. Saudi Arabia alone will produce more than 50 per cent of the expected total output. Kuwait and the UAE will each produce over 8 thousand tons per year.

3.4.3 Petrochemicals

Hydrocarbons from petroleum and natural gas account for most of the chemicals produced today. Although it is difficult to devise a simple system of classification to include all petrochemicals, it is customary now to use three broad categories to identify these products; namely, basic, intermediate and final products.

The main petrochemical basic products are the olefins (ethylene, propylene, butadiene), aromatics (benzene, toluene, xylenes) and methanol. Two primary processes are used in their production: steam cracking of naphtha for the olefins and catalytic reforming for the aromatics. A third process — steam reforming — is also used to synthesize ammonia and methanol. These products form the building blocks from which final petrochemical products are made. The processing chains from the basic to the final products are many and complex. However, a few chains dominate: ethylene and propylene are the main inputs in the making of plastics; aromatics in the making of synthetic fibres; butadiene and benzene in the production of rubbers; and methanol (converted into formaldehyde) in the manufacture of adhesives.

The economics of petrochemical production is also complex, involving complex technologies, large minimum efficient scales of production, high rates of product obsolescence, rising feedstock prices, and the dominance of transnational corporations (TNCs) in the supply of petrochemical intermediates and performance products.

GCC countries are, however, in a privileged position when it comes to petrochemical products, given their abundant supplies of hydrocarbons, some of which are still virtually untapped (flared gas), and their abundant financial capital that could be productively invested in petrochemical production. Moreover, the low labour coefficients in petrochemical complexes match well the desire of GCC producers to reduce their reliance on foreign labour. In addition, in processing their own raw materials, GCC countries will increase the proportion of value-added embodied in their resource-based products, diversify the market outlets for their products, and expose themselves to the industrial experience that is necessary for effective diversification of their overall economic structures.

The high rates of product obsolescence, the dominance of trans-national corporations TNCs and the high proportion of cost represented by feedstock, call for a strategy of production in the GCC which concentrates initially on mature products whose markets can be captured by price undercutting. Thus, production of basic and simple intermediate products should precede the production of performance or end products. This does not argue for a total neglect of the end products; rather it argues for a gradual escalation of the complexity of the product structure in step with increasing experience. The historical record of this industry reveals a strong tendency for migration of production across geographical areas. Production started in the USA but migrated to Western Europe, then to Japan, and now to the centrally planned economies. There is nothing to preclude its migration to what might be regarded as its 'natural abode' in the Arabian Gulf. The excess capacity of production in Europe and Japan is economically inefficient and vulnerable. The Arabian Gulf producers could use their strong leverage in world trade – they are large importers – and their position as major oil suppliers to obtain a substantial share of the market.

The high capital costs of petrochemical projects in the GCC region are balanced by the low variable costs of production. Besides, the capital costs themselves could be lessened by increasing the levels of domestic inputs in design, installation and management. The increased commitment of resources to build large-scale complexes increases the credibility and perceived seriousness of the GCC countries in the quest for a share of the world market.

There was a long time-lag between oil production and the development of petrochemical production in the GCC region. Low oil prices, and consequently limited capital, precluded the development of such a capital-intensive industry as petrochemicals. It was not until the late 1960s that petrochemicals were produced in the region. Fertilizers were first produced in Kuwait in 1966, and then at Dammam, in Saudi

Arabia, in 1970. Since then, a large number of factories have been established in the region, producing urea and other types of fertilizers, and further development is under way. Table 3.12 presents figures relating to the existing and planned capacity for fertilizer production in the GCC region.

Qatar was the first GCC country to embark on the production of basic petrochemical products, with an ethylene complex in the industrial zone at Umm Said in 1974. The complex had a capacity of 280 thousand tons per year of ethylene and 140 thousand tons per year of LDPE. It was further expanded in 1980 to allow the production of 70 thousand tons per year of HDPE.

There is currently no production of petrochemicals other than fertilizers in Kuwait. However, as Table 3.13 shows, a number of projects are contemplated. Similarly, the UAE is also planning a number of new projects, but there is none on-line yet. Ethylene production is the main product planned by the Abu Dhabi National Oil Company. In Kuwait, however, a whole range of basic, intermediate and final products are being studied (e.g. ethylene, HDPE, ethylene glycol, styrene, ortho-xylene and para-xylene).

A joint venture is under construction in Bahrain to produce ammonia and methanol, with planned production of 1,000 tons per day of each scheduled for 1984. Saudi Arabia, Bahrain and Kuwait are co-operating in this venture. There are no other petrochemical projects in Bahrain. Oman is currently studying the feasibility of producing ammonia and urea.

The largest regional petrochemical complexes are planned for Saudi Arabia, and two large industrial cities are under construction to accommodate them. The Saudi Arabia Basic Industries Corporation (SABIC) is entrusted with operating these complexes, together with a number of TNCs. By 1986, the aggregate productive capacity is expected to reach 1.5 million tons per year of ethylene, 660 thousand tons of LDPE and 1.25 million tons of methanol, together with some small quantities of other products (see Table 3.13).

By the mid-1980s, the combined capacity of production of petrochemicals in the GCC countries will include the following: about 2.7 million tons per year of ethylene, or about 5.7 per cent of expected world production; 1.6 million tons per year of methanol, or 9.5 per cent of world production; 280 thousand tons per year of ethanol, or 7 per cent of world production; 655 thousand tons per year of ethylene glycol, or 12.1 per cent of world production; 635 thousand tons per year of styrene, or 5.4 per cent of world production; 800

thousand tons per year of LDPE, or 5.1 per cent of world production; 400 thousand tons per year of HDPE, or 4.9 per cent of world production; and 3.2 million tons per year of ammonia, or 6.8 per cent of world production (see Tables 3.12 and 3.13).

These shares are not high, and certainly far below the shares of proven reserves of gas and oil in the region and the corresponding production and export shares. The restriction of output to a narrow range of products is a wise short-term decision; broadening the base should be tied to the gaining of marketing experience.

Equally important is the linking of investments abroad to GCC exports. This has already happened, but further and immediate attention is called for. The current projects include: Kuwait, for example, owns 40 per cent of the Turkish Mediterranean Petrochemical Company; the Kuwaiti Fund is financing a urea and ammonia complex in Sri Lanka; Qatar owns 40 per cent of the French North Company which operates a petrochemical complex in France; Saudi Arabia has a petrochemical joint venture in Pakistan and another is contemplated with India.

3.5 Conclusion

The GCC is the largest oil producer in the world with the largest pool of proven reserves. What is not so often appreciated is that the region also has large gas reserves. The abundance of oil and gas in the GCC region is, none the less, a temporary phenomenon, as these resources are non-renewable and will eventually run out.

Moving 'downstream' to integrate the various phases and sequences of production — refining, processing, marketing — has become a dominant strategy for the region. Upgrading the domestic value-added component for resource-based products, providing a productive vent for their financial surpluses, diversification of the markets for their products, vertical integration and control over the basic resources, form the essential goals of a policy of industrial development.

The choice of mature products where the highest fraction of production cost is the value of raw materials, and where vulnerability to technological obsolescence is least, is a wise initial choice. The commitment of large funds to large complexes is an intelligent and 'credible threat' strategy to persuade competitors of the seriousness of the Gulf producers' intention to penetrate world markets in these products, and hence to back off themselves. The choice of joint ventures with TNCs should be complemented, perhaps on a larger and broader scale, with joint processing complexes in the Third World and particularly in the Arab world.

The range of petrochemical products should be narrow for the short term, but this need not be the case for the long term. The production of aromatics is one of the more promising possibilities in the shorter term; there is a marked absence of synthetic fibres production in the Arab world that is totally unjustifiable.

A strategy of collective action may be effective in obtaining an increased share of world production for the GCC region. Competition among the GCC countries is harmful to all and rationalization of production calls for a common strategy. Size is a strategic variable in the petrochemical industry, not only from the point of view of scale economies, but also in terms of power in world markets. This industry has a natural linkage to the resource base of the region; raw material costs exceed 75 per cent of total production costs in the case of most of the products and represent a large share of total costs generally for the industry as a whole. The credibility of the GCC in its stated intention to capture its share of world markets increases with the amount of capital committed and with the perceived extent of co-operation and collective action by the countries of the region.

Table 3.1: GCC Proven Crude Oil Reserves, 1970-1980 (millions of barrels)

	1970	1971	1972	1973	1974	1975	1976	1977	1978	1979	1980
Kuwait	79,950	78,198	72,900	72,750	81,450	71,200	70,550	70,100	69,440	68,530	67,930
Qatar	4,300	6,000	7,000	6,500	6,000	5,850	5,700	5,600	4,000	3,760	3,585
Saudi Arabia	141,350	157,475	146,000	140,750	173,150	151,800	153,150	153,100	168,940	166,480	168,030
UAE	12,783	20,502	22,768	25,500	33,920	32,200	31,200	32,425	31,316	29,411	30,410
Oman[a]	1,100	1,000	1,500	1,700	1,600	1,500	1,500	1,800	2,000	2,300	2,400
Bahrain[a]	330	360	380	360	330	310	311	290	270	251	233
GCC	239,813	258,135	250,548	247,560	296,450	262,860	262,411	263,315	275,966	270,732	272,588
OPEC	412,431	430,983	428,373	421,815	484,970	449,870	438,995	439,915	444,936	435,591	434,355
World	611,398	631,856	666,883	627,857	715,697	658,686	636,990	645,848	641,608	641,624	648,525
OPEC/World (%)	67.5	68.2	64.2	67.2	67.8	68.3	68.9	68.1	69.3	67.9	67.0
GCC/OPEC (%)	58.1	59.9	58.5	58.7	61.1	58.4	59.8	60.0	62.0	62.2	62.8
GCC/World (%)	39.2	40.8	37.6	39.4	41.4	39.9	41.2	40.8	43.0	42.2	42.0

Note: a. Oman and Bahrain are not members of OPEC.
Source: *Oil and Gas Journal*, various issues; in the cases of Oman and Bahrain, data are taken from *World Production and Reserve Statistics, Oil and Natural Gas, 1980* (Petroconsultants SA).

Table 3.2: GCC Crude Oil Production, 1970–1980 (thousands of barrels per day)

	1970	1971	1972	1973	1974	1975	1976	1977	1978	1979	1980
Kuwait	2,990	3,197	3,283	3,020	2,546	2,084	2,145	1,969	2,131	2,500	1,664
Qatar	362	431	482	570	518	438	497	445	487	508	471
Saudi Arabia	3,799	4,769	6,016	7,596	8,480	7,075	8,577	9,200	8,301	9,533	9,901
UAE	800	1,060	1,203	1,533	1,679	1,664	1,936	1,999	1,831	1,831	1,702
Oman[a]	322	294	281	293	291	342	366	340	314	295	283
Bahrain[a]	77	75	70	68	67	61	58	58	55	51	48
GCC	8,350	9,825	11,335	13,081	13,581	11,664	13,580	14,010	13,119	14,718	14,069
OPEC	23,413	25,326	27,094	30,989	30,729	27,155	30,738	31,253	29,805	30,929	26,878
World	45,720	48,219	50,850	55,803	56,088	53,384	57,883	59,862	60,143	62,747	59,740
OPEC/World (%)	51.2	52.5	53.3	55.5	54.8	50.9	53.1	52.2	49.6	49.3	45.0
GCC/OPEC (%)	35.7	38.8	41.8	42.2	44.2	43.0	44.2	44.8	44.0	47.6	52.3
GCC/World (%)	18.3	20.4	22.3	23.4	24.2	21.8	23.5	23.4	21.8	23.5	23.5

Notes: Figures in this table may not add to totals or subtotals because of rounding.
a. Oman and Bahrain are not members of OPEC.
Source: OPEC, *Annual Statistical Bulletin, 1980*; MacNaughton, *Twentieth Century Petroleum Statistics*, Petroleum Intelligence Weekly (1980); in the cases of Oman and Bahrain, data are taken from *World Production and Reserve Statistics*, *Oil and Natural Gas, 1980* (Petroconsultants SA).

Table 3.3: GCC Crude Oil Exports, 1970-1980 (thousands of barrels per day)

	1970	1971	1972	1973	1974	1975	1976	1977	1978	1979	1980
Kuwait	2,580	2,775	2,925	2,642	2,203	1,803	1,791	1,625	1,761	2,083	1,297
Qatar	363	483	482	570	511	428	487	410	480	495	466
Saudi Arabia	3,217	4,187	5,444	7,015	7,922	6,601	8,032	8,606	7,706	8,818	9,223
UAE	777	1,055	1,203	1,522	1,690	1,661	1,933	1,990	1,816	1,805	1,697
Oman	332	291	280	292	290	342	368	334	316	295	n.a.
Bahrain	—	—	—	—	—	—	—	—	—	—	—
GCC	7,269	8,791	10,334	12,041	12,616	10,836	12,610	12,968	12,079	13,496	n.a.
OPEC	20,223	22,032	24,079	27,547	27,259	24,064	27,462	27,641	26,089	26,839	22,889
World	23,436	25,541	27,954	31,569	31,344	28,519	32,086	32,315	31,273	33,836	30,617
OPEC/World (%)	86.3	86.3	86.1	87.3	87.0	84.4	85.6	85.5	83.4	79.3	74.8
GCC/OPEC (%)	35.9	39.9	42.9	43.7	46.3	41.6	45.9	46.9	46.3	50.3	n.a.
GCC/World (%)	31.0	34.4	37.0	38.1	40.3	38.0	39.3	40.1	38.6	40.0	n.a.

Note: Figures in this table may not add to totals or subtotals because of rounding.
Source: OPEC, *Annual Statistical Bulletin*, various issues; US Department of Energy, *International Petroleum Annual*, various issues.

Table 3.4: Production and Utilization of Natural Gas in the GCC Region: Selected Years (millions of cubic metres per year)

	1971			1975			1980		
	Production	Flared	% Waste	Production	Flared	% Waste	Production	Flared	% Waste
Kuwait	18,228	11,979	65.7	11,208	4,310	38.5	8,780	1,416	16.1
Qatar	4,515	3,509	77.7	4,730	3,524	68.8	6,400	1,190	18.6
Saudi Arabia	25,481	19,896	78.1	47,230	37,412	79.2	53,265	38,368	72.0
UAE (Abu Dhabi)	10,430	9,385	90.0	14,309	12,938	90.4	14,859	7,608	51.2
Bahrain	506	n.a.	n.a.	3,043	n.a.	n.a.	3,715	509	13.7
GCC	59,159	44,769	75.6	80,520	58,184	72.3	87,019	49,091	56.4

Note: Comparable data were not available for Oman; data for the UAE include only Abu Dhabi.
Source: OPEC, *Annual Statistical Bulletin*, various issues; in the case of Bahrain, data are taken from GOIC, *Petrochemical Industries in the Arabian Gulf*, 1980, p. 37 and OAPEC, *Annual Statistical Report*, 1982.

Table 3.5: GCC Gross Production of Natural Gas, 1971–1980 (millions of cubic metres per year)

	1971	1972	1973	1974	1975	1976	1977	1978	1979	1980
Kuwait	18,228	18,344	16,454	13,222	10,827	11,208	10,272	11,124	13,035	8,780
Qatar	4,514	5,097	6,213	5,151	5,437	4,730	4,290	4,650	6,677	6,400
Saudi Arabia	25,481	32,568	33,292	47,310	37,812	47,230	48,700	43,748	50,561	53,265
UAE	10,430	11,215	13,690	13,054	12,233	14,309	15,341	13,553	13,700	14,859
Oman	n.a.	n.a.	n.a.	n.a.	n.a.	n.a.	n.a.	n.a.	n.a.	n.a.
Bahrain	506	1,132	1,602	1,975	2,876	3,043	3,432	3,715	3,715	3,715
GCC	59,159	68,356	71,251	80,712	69,185	80,520	82,035	76,790	87,688	87,019
OPEC	197,221	213,633	251,964	250,794	222,217	253,706	268,611	268,327	302,618	270,412
World	1,179,000	1,239,000	1,309,000	1,346,000	1,349,000	1,380,000	1,436,000	1,477,000	1,565,000	1,565,000
OPEC/World (%)	16.7	17.2	19.2	18.6	16.5	18.4	18.7	18.2	19.3	17.2
GCC/OPEC (%)	30.0	32.0	28.3	32.2	31.1	31.7	30.5	28.6	28.9	32.2
GCC/World (%)	5.0	5.5	5.4	6.0	5.1	5.8	5.7	5.2	5.6	5.6

Source: OPEC, *Annual Statistical Bulletin*, various issues; US Department of Energy, *World Natural Gas Annual*; Institute of Geological Sciences, *World Mineral Statistics*, various issues; GOIC, *Petrochemical Industries in the Arabian Gulf*, November 1980, p. 37.

Table 3.6: GCC Proven Natural Gas Reserves, 1971-1980 (billions of cubic metres)

	1971	1972	1973	1974	1975	1976	1977	1978	1979	1980
Kuwait	909	856	852	834	829	791	787	782	775	768
Qatar	182	182	182	182	171	625	909	909	1,364	1,364
Saudi Arabia	1,295	1,243	1,270	1,356	2,447	1,503	2,003	2,205	2,190	2,568
UAE	250	295	341	464	522	511	489	491	466	472
Oman	46	46	46	49	46	46	46	46	46	46
Bahrain	93	93	93	153	127	69	69	162	208	208
GCC	2,774	2,714	2,784	3,036	4,142	3,546	4,303	4,595	5,049	5,426
OPEC	12,617	12,606	14,938	23,120	18,362	18,107	23,565	23,380	23,870	24,064
World	47,350	52,214	56,487	68,817	61,787	63,259	69,247	69,342	71,320	73,462
OPEC/World (%)	26.6	24.1	26.4	33.6	29.7	28.6	34.0	33.7	33.5	32.8
GCC/OPEC (%)	22.0	21.5	18.6	13.1	22.6	19.6	18.3	19.6	21.2	22.5
GCC/World (%)	5.9	5.2	4.9	4.4	6.7	5.6	6.2	6.6	7.0	7.4

Note: Figures in this table may not add to totals or subtotals because of rounding.
Source: *Oil and Gas Journal*, various issues.

Table 3.7: Average Chemical Structure of Associated Gas in Selected GCC Countries

Compound	% Volume			
	Bahrain (offshore fields)	Qatar	Ghawar, Saudi Arabia	UAE (Abu Dhabi)
Methane	70.2	55.5	51.0	55.7
Ethane	6.6	13.3	18.5	16.6
Propane	4.5	9.7	11.5	11.6
Iso-butane	0.7	1.6	—	1.6
Butane	2.0	4.0	4.4	3.8
Pentane	1.6	2.6	1.6	—
Hexane	0.9	1.1	0.4	0.7
Heptane and heavier	0.5	1.2	0.2	0.9
Carbon dioxide	4.6	7.0	9.7	5.6
Hydrogen sulphide	—	2.9	2.2	0.8
Nitrogen	8.3	1.1	0.5	0.6
Other	—	—	—	—
Total	100.0	100.0	100.0	100.0

Note: Figures in this table may not add to totals because of rounding.
Source: *Oil and Gas Journal*, various issues.

Table 3.8: Average Chemical Structure of Natural Gas (Dry) in Selected GCC Countries

Compound	Bahrain	% Volume Qatar (Gas Khaf)	UAE (Dubai-Rashid)
Methane	80.0	80.0	78.6
Ethane	1.7	1.3	9.2
Propane	0.4	0.4	3.8
Iso-butane	0.1	0.1	0.6
Butane	–	0.1	1.3
Pentane	0.1	0.2	0.9
Hexane	0.1	–	0.5
Heptane and heavier	–	0.1	0.7
Carbon dioxide	6.6	4.4	3.7
Hydrogen sulphide	–	0.2	0.4
Nitrogen	10.9	13.6	0.3
Other	–	–	–
Total	100.0	100.0	100.0

Note: Figures in this table may not add to totals because of rounding.
Source: *Oil and Gas Journal*, various issues.

Table 3.9: Oil Refining Capacity in the Arabian Gulf (thousands of barrels per calendar day at year end)

	Existing Capacity 1981	Firm Capacity 1986	Firm Capacity Increase 1981–1986	Additional Possible Capacity[a] 1981–1986	Firm + Possible Capacity Increase 1981–1986	Total Firm + Possible Capacity 1986
All Capacity						
UAE	15	195	180	550	730	745
Bahrain	250	250	—	—	—	250
Saudi Arabia	644	2,234	1,590	53	1,643	2,287
Oman	—	50	50	—	50	50
Qatar	11	61	50	—	50	61
Kuwait	594	594	—	106	106	700
Total	1,514	3,384	1,870	759	2,579	4,093
Estimated Export Capacity[b]						
UAE	—	100	100	550	650	650
Bahrain	225	200	(25)	—	(25)	200
Saudi Arabia	300	1,365	1,065	53	1,118	1,418
Oman	—	—	—	—	—	—
Qatar	—	—	—	—	—	—
Kuwait	475	475	—	106	106	581
Total	1,000	2,140	1,140	709	1,849	2,849

Notes: a. 1981 onstream + under construction + committed (e.g. contract let as of mid-January 1981).
b. Estimated on basis of announced intentions and domestic market growth expectations.
— Figures in parentheses indicate decreases.
Source: Based on Field Missions and Reports of GOIC.

Table 3.10: Oil Refineries in the GCC Region, 1980

	Location	Capacity b/d	Type
Kuwait	Mina Abdallah	120,000	D
	Mina Saoud	50,000	D/R/B
	Shuaiba	180,000	D/H/R/C
	Mina Abdallah	120,000	D/H
	Mina Ahmadi	250,000	D/R/B
	Mina Ahmadi (under construction)	250,000	H/L
Saudi Arabia	Ras Tanura	500,000	D/R/B
	Jeddah	70,000	D/R/B/H/VIS
	Riyadh	20,000	D/H/VIS/R
	Ras Tanura (under construction)	25,000	R
	Jeddah (expansion)	170,000	D/VIS/R
	Al-Jubail (under construction)	120,000	D
	Yanbu (under construction)	250,000	D
	Al-Jubail (under construction)	250,000	D
	Riyadh (expansion)	120,000	D/R/H
	Rabgh (under construction)	350,000	D
Bahrain	Awali	250,000	D/C/R/VIS/B
Qatar	Umm Said	6,321	D/R
	Umm Said (planned)	50,000	D/R
UAE	Umm Al-Naar	15,000	D/R/H
	Al-Ruwais	120,000	D/R/H
	Jebel Ali	200,000	D/R/H
Oman	(under construction)	50,000	–

Key: B = Bitumen, C = Cracking, D = Distillation, H = Hydrocracking, L = Lubricating oil, R = Reforming, VIS = Visbreaking.
Source: GOIC, *Petrochemical Industries in the Arabian Gulf*, November 1980, p. 42.

Table 3.11: Liquid Natural Gas Projects in the GCC Region

	Location	Present Status	Feedstock MCF/day	Products (thousand tons/year)					Total Liquid Gases
				Ethane	Propane	Butane	Natural Gasoline	LNG	
Kuwait	Mina Ahmadi	Operational	554	—	556	560	476	—	1,592
	Shuaiba	Operational	1,680	—	3,176	1,717	1,716	—	6,609
	Sub-total		2,234	—	3,732	2,277	2,192	—	8,201
Saudi Arabia	Ras Tanura	Operational	1,000	—	3,500	3,000	3,000	—	9,500
	Juaima	Under construction	—	1,423	2,851	2,190	1,914	—	8,378
	Yanbu	Under construction	3,000	1,262	2,376	1,200	990	—	5,828
	Sub-total		—	2,685	8,727	6,390	5,904	—	23,706
Bahrain	Manama	Operational	100	—	80	75	125	—	280
Qatar	Umm Said	Operational	360	—	336	270	270	—	876
	Umm Said	Under construction	340	—	270	157	113	—	540
	Sub-total		700	—	606	427	383	—	1,416

Table 3.11: Liquid Natural Gas Projects in the GCC Region *(contd)*

UAE	Das Island	Operational	550	—	650	420	220	2,300	3,590
	Al-Ruwais	Under construction	913	—	950	1,426	2,138	—	4,514
	Jebel Ali	Operational	140	—	311	222	244	—	777
	Sub-total		1,603	—	1,911	2,068	2,602	2,300	8,881
GCC Region	—	Operational	—	—	8,609	6,264	6,051	2,300	23,224
		Under construction	—	2,685	6,447	4,973	5,155	—	19,260
	Total		—	2,685	15,056	11,237	11,206	2,300	42,484

Note: MCF stands for millions of cubic feet.
Source: GOIC, *Petrochemical Industries in the Arabian Gulf*, November 1980, p. 40.

Table 3.12: Existing and Planned Fertilizer Industries in the GCC Region

	Location	Products	Capacity Thousand tons/year	Status
Kuwait				
Petrochemical Industries Corporation	Shuaiba	Ammonia	660	Operating since 1966
	Shuaiba	Urea	792	Operating since 1971
	Shuaiba	Ammonium sulphate	165	Operating since 1971
	Shuaiba	Ammonia	330	Planned
Saudi Arabia				
SAFCO	Dammam	Ammonia	180	Operating since 1970
	Dammam	Urea	300	Operating since 1970
SEMAD	Al-Jubail	Ammonia	330	Operational in 1983
	Al-Jubail	Urea	500	
	Al-Jubail	Ammonia	330	Under study
	Al-Jubail	Urea	500	
Bahrain				
Gulf Petrochemical Corporation (joint venture)	Satrat	Ammonia	330	Operational in 1984
Qatar				
QAFCO	Umm Said	Ammonia	297	Operating since 1973
	Umm Said	Urea	330	Operating since 1973
	Umm Said	Ammonia	297	Operating since 1979
	Umm Said	Urea	330	Operating since 1979
UAE				
ADNOC	Al-Ruwais	Ammonia	330	Operational in 1983
	Al-Ruwais	Urea	500	Operational in 1983

Table 3.12: Existing and Planned Fertilizer Industries in the GCC Region *(contd)*

Oman				
Ministry of Oil and Mineral Resources	Sahar	Ammonia	200	Planned
		Urea	330	Planned
GCC Region		Ammonia	1,434	Operating
		Urea	1,752	Operating
		Ammonia	990	Operational in 1983–4
		Urea	1,000	Operational in 1983–4
		Ammonia	860	Planned or under study
		Urea	830	Planned or under study
		Ammonia	3,284	Total
		Urea	3,582	Total
		Ammonium sulphate	165	Total

Source: GOIC, *Petrochemical Industries in the Arabian Gulf*, November 1980, p. 48.

Table 3.13: Existing and Planned Petrochemical Projects in the GCC Region

	Location	Products	Capacity Thousand tons/year	Status
Kuwait				
KPIC	Shuaiba	Ethylene	350	Planned
		HDPE	130	Planned
		Ethylene glycol	135	Planned
		Styrene	340	Planned
		Benzene	280	Planned
		Ortho-xylene	60	Planned
		Para-xylene	86	Planned
Saudi Arabia				
(a) Saudi Petrochemical Co (Shell Oil Co.)	Al-Jubail	Ethylene	656	
		Ethylene dichloride	456	
		Styrene	295	Operational in 1985
		Ethanol	281	
		Caustic soda	377	
(b) Saudi Yanbu Petrochemical Co. (Mobil Chemical Co.)	Yanbu	Ethylene	450	
		LDPE	200	Operational in 1985
		HDPE	90	
		Ethylene glycol	220	
(c) Al-Jubail Petrochemical Co. (Exxon Chemical Co.)	Al-Jubail	LDPE	260	Operational in 1985
(d) Saudi Methanol Co. (Japanese consortium)	Al-Jubail	Methanol	600	Operational in 1983

Table 3.13: Existing and Planned Petrochemical Projects in the GCC Region *(contd)*

(e) National Methanol Co. (Celanese-TEXAS Eastern)	Al-Jubail	Methanol	650	Operational in 1985
(f) Arabian Petrochemical Co. (Dow Chemical Co.)	Al-Jubail	Ethylene LDPE HDPE	500 70 110	Operational in 1985
(g) Eastern Petrochemical (Japanese consortium)	Al-Jubail	LDPE Ethylene glycol	130 300	Operational in 1985
Bahrain				
Gulf Petrochemical Industries jointly with Kuwait and Saudi Arabia	Satrat	Methanol	330	Operational in 1984
Qatar				
QAPCO and CDF	Umm Said	Ethylene LDPE Propylene HDPE	280 140 50 70	Operating
UAE				
(ADNOC)	Al-Ruwais	Ethylene	450	Under consideration
GCC Region		Ethylene Ethylene dichloride Ethylene glycol HDPE	2,686 456 655 400	

Table 3.13: Existing and Planned Petrochemical Projects in the GCC Region *(contd)*

GCC Region *(contd)*		
LDPE	800	Total operational, operational in 1985, planned or under study
Styrene	635	
Benzene	280	
Propylene	50	
Ortho-xylene	60	
Para-xylene	86	
Methanol	1,580	

Source: Al-Wattari, *Oil Downstream*, Kuwait, OAPEC, 1980, pp. 98, 99; SABIC, *The Fourth Annual Report for 1400 A.H. (1980 A.D.)*, p. 22.

4 THE NON-OIL SECTORS: THE QUEST FOR DIVERSIFICATION

Since oil reserves are finite and non-renewable and the demand for oil is not stable and continuous, diversification of the productive basis of the regional economies assumes critical importance. This diversification necessitates maintaining a high ration of investment in non-oil productive sectors over a long period. For the Arab Gulf states, high rations of investment in non-oil GDP require little or no sacrifice of present consumption. Saving can accrue principally from the oil sector. However, this relationship has not been without negative consequences. The facts show that the relationship between the growth of oil revenues and the growth of non-oil GDP in the region has been rather tenuous.

In each of the countries of the region, the growth rate of non-oil GDP drops if the factor of oil revenues is taken out of the calculations.[5] This suggests that the other sectors of the economy would not grow on their own without the continuous flow of oil. Oil has not yet succeeded in promoting a state of sustained growth in the non-oil sectors, which are still heavily dependent on oil developments. The weak structures of the non-oil sectors are a consequence of a number of complex factors, among them ironically is the oil boom and the limited productive investments in these sectors.

Two basic features of economic policy in the Arab Gulf states over the last decade go a long way in explaining the relative lack of auto-dynamism in the non-oil sectors: heavy investments in the infrastructure of the economy that were mainly divorced from productive investments, and the sharply negative effects of oil on agriculture.

The heavy emphasis on the development of infrastructure does not mean that investment in industry or agriculture is not taking place, but rather that one type of investment has so far been predominant. The development of infrastructure is certainly necessary for the development of productive sectors, but the two types of investment are not separable.

The development of infrastructure without tying it directly to productive activity vitiates the economic effort in two fundamental ways. First, it raises the average social unit cost of use. Second, such investments are a drain on future capital budgets, as maintenance will eat up over time a large portion of future revenues, leaving less available for other alternatives. Were productive investments made simultaneously, their social surplus might be used to defray such costs. The heavy emphasis on infrastructural development in the region was almost divorced from productive investments and some have even gone as far as to suggest that it had taken place at its expense.

The rise in oil revenues was accompanied by massive increases in food imports into the region. This inflow of subsidized imports depressed food prices domestically to the extent that domestic farmers' opportunity costs became substantially high and many of them left the farms for the cities. The decrease in food prices was matched by very high costs of production and labour, so much so that farmers were caught in a cleft stick – low prices for their products and high costs for their inputs. Although this latter phenomenon was not generalized in the GCC region, and many policies were promulgated recently to deal with it, the mid-1970s were characteristic of this situation, particularly in Oman and parts of Saudi Arabia where agricultural activity is potentially promising.

Equally important in explaining the weak economic structures of the non-oil sectors of the Gulf economies is the limited economic size of each member state. The small size of almost all of the states in the region has contributed substantially to their limited absorptive capacities and, therefore, to their present lopsided structures of production. The smallness of the domestic economy constrains the prospects for balanced growth not only in terms of the limited size of the market, but also in terms of paucity of resources, technological flows and diversity of opportunities. Although the GCC region is generally homogeneous culturally, politically and economically, it is geographically differentiated. Agriculture, for instance, is not equally viable everywhere; Saudi Arabia and Oman appear to offer the only credible potential for agricultural development.

More important perhaps is the fact that some resources are common resources to the region. In other words, if one state were to use more of any of these resources it reduces what is available to the rest of the region. This is particularly true in the case of water and fishing. In such instances, only regional co-ordinated exploitation could result in acceptable and efficient arrangements for tapping these resources.

Lopsided economic structures are the outcome of historical processes and as such are symptomatic of dynamic disequilibrium. The process feeds on itself and needs to be reversed in a massive and decisive way. When an economy is small, it cannot absorb easily productive investment and as such it remains a small market that does not induce further investments. If each state in the region is in a similar bind, then all will obviously have similar structures (empty production boxes) and will have little to trade and as such the region as a whole will be economically circumscribed. Thus the non-complementary structures of the regional economy are natural outcomes of limited size and lack of co-ordinated development. It follows that co-ordinated investment and pooling of markets and resources could alter this dynamic process. Togetherness and the harmonization of development are necessities of the present economic reality in the Gulf; without them the region will remain locked in a state of competitive and weak structures, conflicting interests and limited potential.

The drift of the argument above is simple, the small economic size of each state precludes the possibility of meaningful diversification and balanced development. It precludes the attainment of security and self-reliance and acts as a dynamic process that will continue to prevent the development of regional complementarity. Integration of the regional economies is a natural economic imperative of the dynamics of balanced growth. The analysis to follow of the prospects of diversification of the regional economies will, therefore, be conducted at the only feasible level — the region as a whole.

Two major sectors will be singled out for discussion — agriculture (Chapter Five) and non-oil manufacturing (Chapter Six). Indeed, there are many other activities to be considered and each with its own rationale for inclusion, but the chosen subsets dominate in importance and prospects and that is why they were chosen for special emphasis.

The economy operates as an integrated network and, therefore, the singling out of component sectors for analysis is only artificial and must be seen in this light.

5 AGRICULTURAL DEVELOPMENT IN THE GCC REGION: IS IT POSSIBLE?

5.1 Introduction[6]

Agriculture is a crucial sector in the countries of the Gulf Co-operation Council, not because of its ability to generate high levels of domestic or foreign earnings, but because of its potentially limiting influence on balanced development. Commodity-producing sectors in most economies are the major contributors to income and employment. The more diversified an economy is, the more secure its income and employment opportunities and the more extensive are its prospects for balancing its development. In this respect, agriculture plays a critical role in each and every aspect of self-reliance, food security, diversification and balanced development.

Agriculture in the GCC states exhibits all of the characteristics of dualism. It can be partitioned into a highly mechanized commercial subsector and a traditional subsistence-type subsector. Thus, agriculture presents within itself the same development problems as the total economy of the region. This chapter examines the agricultural resources of the various Gulf states. The objective of the examination is to place agriculture in a development context for the regional economy. To this end the particular strengths and weaknesses of agriculture are examined. Demand conditions for agricultural products are noted, as are supply capabilities. Existing forward and backward linkages are presented with a view to setting the context for further resource-based processing within the region.

Linkages among agriculture and the other sectors of the economy are not well documented. Figure 5.1 (see p. 86) provides an indication of these macro or sectoral linkages. It depicts the major links between agriculture and the local suppliers to and purchasers from the agricultural sector. The economic linkages to other primary sectors are strong. Fisheries, oil and gas, water and mineral sectors all have links to agriculture. The nature of the various linkages is discussed here to highlight the development strategy in this sector.

5.2 The Agricultural Base of the Region

Fundamentally, the GCC region is not well suited to agriculture. Although agricultural production is a traditional economic activity of the population under natural circumstances, the climate and land severely restrict both the types and quantities of crops and livestock. Table 5.1 (see p. 87) indicates the land base of the six countries in terms of total area and agricultural land of various types. The crucial element in this table is the relative scarcity of agricultural land in the region. Of the total land base of 247,519,000 hectares only 1,164,000 hectares or 0.47 per cent is suitable for crops and 87,522,000 hectares or 35.37 per cent is suitable for permanent pasture. The extreme scarcity of land suitable for cultivation places a very real limit on the quantity of crops that can be grown, even with massive infusions of such inputs as labour, machinery, chemicals and water. Table 5.1 also indicates that Saudi Arabia dominates agricultural production in the region. This dominance arises from its relative size, but also because Saudi Arabia and Oman are the only two countries in the region that have areas with sufficient precipitation to allow agricultural production without irrigation. Those areas are limited in size but constitute a significant resource.

Irrigation is the key factor in developing increased agricultural production in the GCC region. As Table 5.1 indicates, although irrigated acreage is increasing in Saudi Arabia and Oman it has been relatively stable in the other countries over the past decade. The Food and Agriculture Organization (FAO) estimates of irrigated land are necessarily somewhat crude, owing to the generally limited data available from the respective countries. However, even if irrigated land is expanding at a faster rate than is indicated in Table 5.1, the potential land base in Bahrain, Kuwait and Qatar will soon be exhausted. Thus, if the GCC seeks to expand cropland it will have to be in Saudi Arabia and Oman or, to a lesser extent, the United Arab Emirates.

Pasture figures tend to exaggerate the relative potential for livestock herds. Even for livestock species that are well adapted to the indigenous vegetation, such as camels, goats and sheep, the animal density rates on native pasture are very low. With the exception of southern Oman, pasture conditions without irrigation in the region are poor. Only in the immediate vicinity of oases does natural pasture provide sufficient plant matter to sustain continuous grazing. Even then only a limited number of animals can be supported.

Despite the relative paucity of land resources and the harsh climate, a significant portion of the population has been engaged, historically, in agriculture within the region particularly in Oman and Saudi Arabia. Although this proportion is continually falling as the countries become more urbanized, agriculture is still the dominant source of employment in Saudi Arabia and Oman. Table 5.2 indicates the relative proportions of the population employed in agriculture (where the figures are available). In the four smaller countries, the proportion of the population in agriculture is small, reflecting the dominance of major urban centres. Note too that in addition to the slow decline in the share of population in the agricultural sector, there is also a decline in the economically active group as a proportion of the total agricultural population. These figures are consistent with the movement of younger individuals lured to urban centres by better opportunities, leaving the rural areas with an ageing population structure.

Production capabilities of those engaged in agriculture are exceedingly divergent. At one end of the spectrum are those employed at the large turn-key livestock projects where the most modern technology is employed. At the other end are nomadic herdsmen whose production technology is hundreds of years old. There are two fundamental labour problems in agriculture: the first is how to reduce employment in the agricultural sector to man the emerging activities in other sectors; the second is how to improve the skills of those remaining. The same impediments limit solutions to both problems. Those engaged in traditional agriculture have no skills other than traditional farming skills and limited opportunities to acquire new skills. Thus, individuals lack the basic skills to participate in the expanding sectors of the economy. At the same time, they also lack the skills required to adapt to high-technology agriculture.[7]

Structural problems also inhibit the expansion of production. For traditional farmers, average farm holdings are far less than the minimum size necessary to operate a viable modern family farm. Similarly, adoption of modern technology is limited by scarcity of machinery, spare parts and mechanics to service the equipment. The various countries are attempting to alleviate these problems by subsidizing domestic production, providing concessionary financing for inputs and expanding research and extension programmes. All these programmes are long-term solutions that have limited short-run impacts if farmers are unwilling or unable to adjust production techniques, and the manpower to implement the programmes is in scarce supply.

Various FAO studies and domestic reports note the critical shortage of skilled personnel working in agriculture. These shortages occur in such diverse areas as programme implementation and administration within the ministries themselves; research and extension activities in the various stations; and the supply of skilled technicians on the farm. Low educational levels and skills of the local population make the successful adoption of advanced agricultural production techniques problematic. However, if output is to be expanded, adoption of highly sophisticated irrigation practices and livestock management methods will be necessary, particularly in the adverse climatic conditions such as they are experienced in the Arabian Peninsula. Thus, the crucial requirement for the successful introduction of the advanced agricultural technology that is essential if the GCC states are to increase output is the availability of skilled manpower necessary to operate and maintain the equipment.

5.3 Agricultural Production in the Region

In all six countries of the Gulf Co-operation Council there have been major expenditures to stimulate agricultural production. These efforts have been targeted at expanding traditional crops and cultivation practices and introducing new crops and technologies. Consequently, output has increased significantly in all countries. For certain crops, production meets domestic needs for at least part of the year, and in some instances there are exportable surpluses. However, despite the significant increase in production, demand has increased at a far greater rate, leading in many instances to growing gaps between domestic supply and demand. Table 5.3 indicates the value of domestic agricultural production (including fisheries) in the various GCC countries over the period 1970–81. Over this period, the nominal value of output rose to levels ranging from 3.3 to 40.7 times the 1970 base values. The nominal value of output for the region as a whole increased 7.6 times, from US$291.1 million to US$2,210.9 million.

Information on volumes and values of production by particular crop or livestock type is not readily available, particularly in a consistent form. However, Tables 5.4–5.7 provide rough indicators of the types and relative importance of various crops and domestic animals. The tables show that there has been a considerable increase in the production of fruit and vegetables, particularly tender crops such as tomatoes. Tree fruit production in the region is primarily composed of dates and lemons, and the data suggest that both products are experiencing increases in production. Livestock numbers indicate a significant availability of domestic meat and animal products. The livestock table does not indicate the major source of meat, which is confinement poultry production. Saudi Arabia alone has the capacity to produce over ten million birds per annum (as reported by the Saudi Arabian Ministry of Industry and Commerce).

Yield figures, where available, suggest that output per unit of land is roughly comparable for all countries. This can be attributed to the need for water to sustain agricultural production. Where adequate water supplies are available, yields are roughly the same. Increases in output in the future will be limited by two factors, water availability and land availability. With respect to these two constraints, Saudi Arabia and Oman have a distinct advantage. They are the two largest countries with considerable quantities of arable land that can be readily improved where now cultivated, or brought into production if now uncropped. In addition, they both possess regions where water is less of a constraint than in other parts of the peninsula. The south-west coast of Saudi Arabia and southern Oman are areas where supplemental irrigation is sufficient to produce a crop. In many instances, the required irrigation water can be obtained by impounding storm run-off. Other areas of the peninsula require full irrigation if a crop is to be produced. They also rely on non-renewable ground water or upon very expensive desalinated water, both of which have high opportunity costs. As a result, the true social cost of production is high.

Fruit, vegetables and some row crops can be grown under full irrigation schemes in quantities large enough to meet a significant share of domestic requirements. This is particularly viable in controlled environment greenhouses. For cereals, oil seeds and coarse grains the potential to do so is much lower. These crops have far lower value per hectare, so that water has to be provided at a very low price in order to make the crops profitable. Where the water has a high opportunity cost, its allocation to cereals results in a major waste. In areas where there are soil moisture reserves from precipitation, or surface water is available, irrigation for cereals can be a viable option. Such areas are to be found only in Saudi Arabia and Oman.

As noted frequently above, water availability is crucial for agricultural production in the GCC states. Irrigation water and livestock water are the obvious uses. However, at the hottest time of the year, water is also required to cool plants and animals. Although the water for cooling purposes need not be of good quality, in general there are limits to the quality of water that can be used in agriculture.

Salinity levels in many of the aquifers of the region, however, make them unsuitable for agricultural use without treatment or blending with less saline water. This, however, increases the cost of the water. Once the water is of an acceptable quality for irrigation, problems can still arise.

Over time, with continuous irrigation, salts can build up in the soil leading to decreased fertility, lower yields and, ultimately, zero production. For the most part, this problem can be controlled by appropriate choice of irrigation system and careful management. It would seem, however, given the severe shortage of highly skilled agricultural manpower in the GCC economies, that there is considerable potential for irrigation projects to go wrong. Similarly, the quantity of water necessary for production is highly dependent on the particular technology adopted and the sophistication of the farm management. Drip systems use far less water than traditional flood systems, or even centre pivot systems. They do, however, require higher levels of maintenance and monitoring. Similarly, for flood irrigation more frequent applications can reduce the total quantity of water that must be applied, but in order to do so fields must be levelled and care must be taken in controlling salt build-ups.

5.4 The Developmental Imperatives of Regional Agriculture

There is a growing concern about food security in the region. The individual governments feel that relying virtually completely on imported food products is a dangerous option. There is some fear of an embargo by food-exporting countries, and also a recognition of the difficulty of keeping supply routes open in the area if war breaks out as a result of any of the volatile situations in the region. Consequently, plans to build up a source of domestic supply to reduce this dependency have been put in place.

A second reason for fostering agriculture can be explained by the social structure of the various countries. Agriculture contributes a small and declining share to gross national product, and varying proportions of the countries' citizens are either directly or indirectly employed in agriculture. As noted previously, these people often have limited skills and cannot be readily integrated into the growing high-technology portion of the economy. Therefore, providing funds for agriculture may make sense as a short-run mechanism for distributing income to such people. Investments and subsidies to agriculture can be viewed, from this perspective, as transfer payment mechanisms. Since only nationals can own land, the transfers increase the welfare of citizens but not foreigners. In Bahrain, Kuwait, Qatar and the United Arab Emirates, the proportion of the local population in agriculture is small enough to limit the utility of such transfer mechanisms. However, in Saudi Arabia and Oman, the large agricultural populations can be supported by such payments.

The problem that the GCC countries face — of a growing gap between domestic supply and demand, despite rapidly increasing levels of output — can be attributed to the rapid influx of workers from other countries. To undertake the rapid development of the various economies, large numbers of offshore workers have had to be imported. If the number of foreign workers could be reduced, the demand for food would be reduced accordingly: for example, reducing the number of foreigners by 50 per cent would have an immense impact on aggregate food demands in the various countries of the GCC.

With the general rise in *per capita* income, adjustments in production are required to satisfy growing demand for fruit and vegetables, meat and other superior goods in people's diets. Government incentives of various types are being employed to bring about these adjustments. Thus, the need to change the structure of agriculture provides another reason for government intervention.

With sufficient allocation of resources to agriculture, the region could become self-sufficient. The costs, however, would be astronomical. In order to achieve self-sufficiency, vast amounts of money and fuel would have to be allocated to provide the equipment and water to produce the crops and livestock for consumption. The facts are that, with the notable exception of dates, fish, fruit and vegetables, the levels of sustainable self-supply in the region are rather low. Although the self-supply ratios for eggs, meat and poultry are relatively high, these commodities cannot be produced in volume without feed grains. Thus, without an indigenous source of feed this latter group of commodities could not, realistically, be considered to be in secure supply.

Saudi Arabia has the greatest degree of self-sufficiency. But even for Saudi Arabia, in isolation, the cost of increasing output to meet self-sufficiency requirements would be high. Moreover, the target of self-sufficiency is not fixed. As the population grows, and its wealth increases, demand will also increase. Thus the level of output that may have fed the population of ten years ago with a diet based on a lower level of income will not feed the current population. To achieve self-sufficiency, output must grow at a faster rate than consumption.

Table 5.8 indicates that the share of resources going to agricultural development in the various countries has been declining. It is also true that imports of food and feeds have been growing at a faster rate than imports of machinery. If a policy of self-sufficiency is the objective, these trends will have to be reversed. To do so will impose significant costs on the region, both in direct expenditure and in terms of resource allocation. The major cost in terms of resources will be the production of water for use in agriculture. The cost of this water will be far higher than the world market value of any cereal crops produced using it.

Security of supply is surely an important goal, but at what point do the costs of reaching it exceed the benefits? Security can be obtained in a number of ways. Possible options are long-term purchase agreements from a number of suppliers. These could be used in conjunction with expanded storage facilities to build up buffer stocks. Development of domestic livestock species, such as the camel, to provide meat using rough pasture rather than poultry eating imported grains would reduce the effects of a major impediment to food security. Expansion of domestic production should be considered, but as one of a broad set of objectives if a major misallocation of resources is to be avoided. Equally important are investments in other Arab countries with potential for agriculture – e.g. Sudan; the concept of food security is an Arab phenomenon and must be treated this way.

Although the initial resource base and climatic conditions of the GCC region are not favourable for agricultural production, there is considerable potential for expansion from existing levels of output. Over the past decade, the output of crops and livestock has shown a marked increase. In spite of this increase, net imports have risen because population and income levels have increased at faster rates. The GCC countries have indicated a commitment to expanding the agricultural sector. While this is clearly possible, such an expansion should be approached with care.

Caution is advisable for two reasons. The first is the high opportunity cost of water. Water allocated to agriculture produces outputs that have relatively low market values in world trade, whereas water allocated to other activities can produce higher-value outputs. Where water is expensive, the value of the crops or livestock produced could be less than the value of the water input. Secondly, increases in agricultural production will require major increases in the skill level of the agricultural labour force. Skilled labour is typically in short supply in the GCC region. Without the use of skilled labour, advanced technology can easily destroy the existing resource base. Thus, the costs of increasing agricultural production could be high.

5.5 Co-operation in Agriculture

The existing level of co-operation in agriculture among the member states of the GCC is already quite high. Co-operation occurs on both a formal and an informal basis. At a formal level, there are projects that are jointly sponsored by the six countries, and others that involve the GCC states in co-operation with other Arab countries. At an informal level, the various ministries of agriculture are in contact to facilitate communication and exchange information. These informal contacts provide a potential medium for co-operation, ongoing co-ordination and the development of joint activities. Similarly, special topics are discussed at conferences called for particular purposes. An example is the conference on food security and the food industry, held in Dubai in April of 1981.

Such organizations as the Arab Organization for Agricultural Development (AOAD) and the Council of Arab Economic Unity provide a general forum at the formal level. The AOAD conference on the Arab food reserve stock, which took place in Khartoum in 1980, suggested, for example, that the various Arab countries should hold a three-month reserve stock for security purposes. There are also specialized organizations, such as the Arab Centre of Studies on Arid Zones and Dry Lands, which undertake research on cultivation methods and crops. Similarly, the Arab Company for the Development of Animal Wealth provides funds for livestock development projects.

But more is needed to consolidate the region's agricultural objectives and to realize short-term and long-term targets. Already the governments of the region influence the agricultural sector in a wide variety of ways. These channels need to be rationalized, strengthened and harmonized across the region.

Among the most productive areas of co-operation in agriculture is in the field of agricultural research. The research function is particularly important if the indigenous resources of the countries, in terms of plant and animal species, are to be utilized more effectively. Research to improve yields of native forages for use as livestock feed would reduce the need to import foreign crop varieties and livestock, and the technology to maintain them. In a similar vein, research to adapt foreign crops and livestock to the high heat and high salinity conditions which prevail throughout the largest part of the Arabian Peninsula is an important function. Research in crop production and animal breeding is important, but research in farm management is vital if the biological research is to be successfully implemented. New techniques, new crops and new livestock species would have to be introduced into an existing farm structure. To do so successfully would require considerable knowledge of the existing objectives and methods of farmers and the development and implementation of techniques to incorporate the new with the old.

The line between research in farm management and extension activities is not clear. Extension activities take the results of research and convey them to the farmers in a form which they can make use of. Consequently, extension workers must understand both the farmer and the researcher. In the absence of an effective extension activity, successful adoption of new methods will be slow. If the extension agents can convince the farmer that the benefits from trying something new exceed the costs, the traditional agriculture sector will be able to progress rapidly. If, as is the case for the bulk of the traditional sector in the GCC, skills (including literacy) are low, the role of the individual extension worker will be critical in translating scientific research into readily understood techniques that can be adopted and applied by the farmer.

Water management on a regional basis is also critical. Current practices are wasting this highly scarce resource. There is currently too low a price of water in the region. This encourages excessive use, reducing the supply available for other uses now, and for all uses in the future. Now is the time for working out a careful priority system for water use and a price system capable of rationing use in accordance with the priority system.

5.6 Conclusion

In agriculture, GCC co-operation is a matter of survival. The sooner this co-operation is translated into workable procedures and schemes, the faster will be the development of this sector in the region. Co-operation within the region is only a part of the needed co-operation endeavour. Equally important is co-operation with the rest of the Arab world particularly areas in the Fertile Crescent and

Sudan. Food security is an Arab problem and can only be solved at this level. Besides, balancing sectors in the region will most likely prove possible in the context of Arab economic harmonization.

Figure 5.1: Agricultural Sector Linkages

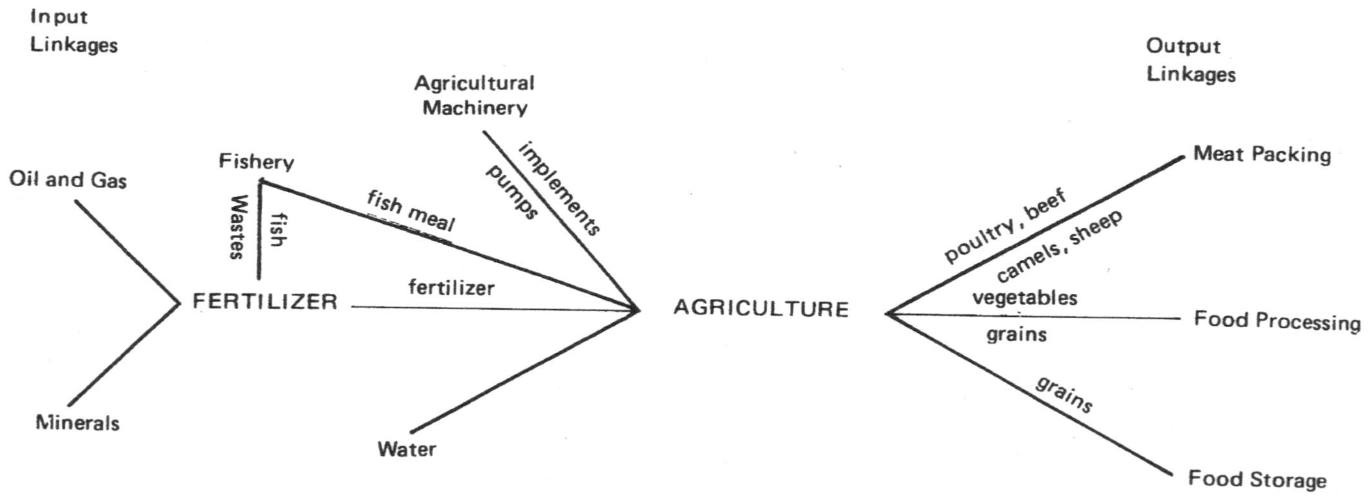

Input Linkages

Output Linkages

Agricultural Machinery — implements, pumps → AGRICULTURE
Fishery — fish meal → FERTILIZER
fish Wastes → FERTILIZER
FERTILIZER — fertilizer → AGRICULTURE
Water → AGRICULTURE
Oil and Gas → FERTILIZER
Minerals → FERTILIZER

AGRICULTURE:
- poultry, beef, camels, sheep → Meat Packing
- vegetables, grains → Food Processing
- grains → Food Storage

Table 5.1: Land Area in the Countries of the GCC Region (thousands of hectares)

	1971	1972	1973	1974	1975	1976	1977	1978	1979	1980
Saudi Arabia										
Land area	214,969	214,969	214,969	214,969	214,969	214,969	214,969	214,969	214,969	214,969
Arable and permanent crops	878	952	1,030	1,115	805	1,110	1,104	1,105	1,105	1,105
Arable land	811	885	960	1,035	730	1,040	1,040	1,040	1,040	1,040
Permanent crops	67	67	70	80	75	70	64	65	65	65
Permanent pasture	85,000	85,000	85,000	85,000	85,000	85,000	85,000	85,000	85,000	85,000
Arable, irrigated	350	360	370	375	390	390	390	395	395	395
Qatar										
Land area	1,100	1,100	1,100	1,100	1,100	1,100	1,100	1,100	1,100	1,100
Arable and permanent crops	2	2	2	2	2	2	2	2	2	2
Arable land	2	2	2	2	2	2	2	2	2	2
Permanent crops	—	—	—	—	—	—	—	—	—	—
Permanent pasture	50	50	50	50	50	50	50	50	50	50
Arable, irrigated	—	—	—	—	—	—	—	—	—	—

Table 5.1: Land Area in the Countries of the GCC Region (thousands of hectares) (contd)

	1971	1972	1973	1974	1975	1976	1977	1978	1979	1980
Oman										
Land area	21,246	21,246	21,246	21,246	21,246	21,246	21,246	21,246	21,246	21,246
Arable and permanent crops	36	36	37	36	36	39	40	36	41	41
Arable land	16	16	16	16	16	17	17	16	18	18
Permanent crops	20	20	21	20	20	22	23	20	23	23
Permanent pasture	1,000	1,000	1,000	1,000	1,000	1,000	1,000	1,000	1,000	1,000
Arable, irrigated	–	–	–	35	–	–	37	–	–	38
Kuwait										
Land area	1,782	1,782	1,782	1,782	1,782	1,782	1,782	1,782	1,782	1,782
Arable and permanent crops	1	1	1	1	1	1	1	1	1	1
Arable land	1	1	1	1	1	1	1	1	1	1
Permanent crops	–	–	–	–	–	–	–	–	–	–
Permanent pasture	134	134	134	134	134	134	134	134	134	134
Arable, irrigated	1	1	1	1	1	1	1	1	1	1

Table 5.1: Land Area in the Countries of the GCC Region (thousands of hectares) (contd)

	1971	1972	1973	1974	1975	1976	1977	1978	1979	1980
Bahrain										
Land area	62	62	62	62	62	62	62	62	62	62
Arable and permanent crops	2	2	2	2	2	2	2	2	2	2
Arable land	1	1	1	1	1	1	1	1	1	1
Permanent crops	1	1	1	1	1	1	1	1	1	1
Permanent pasture	4	4	4	4	4	4	4	4	4	4
Arable, irrigated	1	1	1	1	1	1	1	1	1	1
UAE										
Land area	8,360	8,360	8,360	8,360	8,360	8,360	8,360	8,360	8,360	8,360
Arable and permanent crops	12	12	11	12	12	12	11	12	12	13
Arable land	7	7	6	7	7	7	6	7	7	6
Permanent crops	5	5	5	5	5	5	5	5	5	7
Permanent pasture	200	200	200	200	200	200	200	200	200	200
Arable, irrigated	5	5	5	5	5	5	5	5	5	5

Note: Data on irrigation relate to areas purposely provided with water, including land flooded by river water for crop production or pasture improvement, whether this land is irrigated several times or only once during the year stated. FAO, *Production Yearbook*, 1981, p. 3.

Source: FAO, *Production Yearbook*, various issues, 1976–81.

Table 5.2: Total Population, Agricultural Population and Population Economically Active in Agriculture, in Thousands, for Selected Years, 1970–1980 (blanks indicate data not available)

	1970	1975	1978	1979	1980
Bahrain					
Total population	220	260	280	290	360
Agricultural population					3
Per cent of total population					0.8
Economically active in agriculture					
Per cent of total population					
Per cent of agricultural population					
Economically active					
Kuwait					
Total population	740	1,002	1,215	1,279	1,353
Agricultural population	13	17	21	22	23
Per cent of total population	1.8	1.7	1.7	1.7	1.7
Economically active in agriculture	4	5	6	6	6
Per cent of total population	0.5	0.5	0.5	0.5	0.4
Per cent of agricultural population	30.8	29.4	28.5	27.3	26
Economically active	241	286	329	344	360

Table 5.2: Total Population, Agricultural Population and Population Economically Active in Agriculture, in Thousands, for Selected Years, 1970–1980 (blanks indicate data not available) *(contd)*

	1970	1975	1978	1979	1980
Oman					
Total population	678	766	839	865	891
Agricultural population	442	494	527	538	550
Per cent of total population	65.2	64.5	62.8	62.2	61.7
Economically active in agriculture	120	131	138	139	142
Per cent of total population	17.7	17.1	16.4	16.1	15.9
Per cent of agricultural population	27.1	26.5	26.2	25.8	25.8
Economically active	177	203	218	224	230
Qatar					
Total population	110	170	200	210	220
Agricultural population					
Per cent of total population					
Economically active in agriculture					
Per cent of total population					
Per cent of agricultural population					
Economically active					

Table 5.2: Total Population, Agricultural Population and Population Economically Active in Agriculture, in Thousands, for Selected Years, 1970–1980 (blanks indicate data not available) *(contd)*

	1970	1975	1978	1979	1980
Saudi Arabia					
Total population	6,200	7,180	7,870	8,110	8,370
Agricultural population	4,090	4,534	4,822	4,929	5,033
Per cent of total population	66	63.1	61.3	60.8	60.1
Economically active in agriculture	1,122	1,205	1,258	1,278	1,299
Per cent of total population	18.1	16.8	16.0	15.8	15.5
Per cent of agricultural population					
Economically active	1,699	1,909	2,052	2,104	2,160
UAE					
Total population	230	560	710	750	800
Agricultural population				16	
Per cent of total population				2.1	
Economically active in agriculture		13			
Per cent of total population					
Per cent of agricultural population					
Economically active					

Source: FAO, *Production Yearbook*, 1976–1980; Saudi Arabia, Chamber of Commerce and Industry, *Future Prospects of Economic Integration in the Gulf Cooperation Council Region*, Riyadh, 1982 (Arabic); 'Agriculture and Food Industries in the UAE', *First Symposium on Food Security and Food Industries in the Arab Gulf and Arab Peninsula*, UAE, 1981, p. 16.

Table 5.3: Domestic Agricultural Production in the GCC, 1970-1981 (Agriculture, Forests and Fisheries) (in current US$ million)

	1970	1971	1972	1973	1974	1975	1976	1977	1978	1979	1980	1981
UAE	16.9	20.3	26.4	38.8	52.5	83.1	109.3	125.8	156.0	178.2	223.1	234.6
Bahrain	2.1	2.5	3.0	16.3	19.8	26.3	27.3	34.9	41.6	51.6	73.8	85.4
Saudi Arabia	218.7	235.1	255.4	307.6	350.0	395.0	449.4	529.3	1,149.9	1,248.5	1,397.2	1,648.3
Oman	39.8	40.4	44.3	47.6	50.4	58.5	62.0	69.5	78.5	92.6	107.4	133.8
Qatar	5.4	7.2	9.2	10.8	16.1	19.1	22.9	18.8	27.4	33.2	45.5	46.7
Kuwait	8.2	8.9	11.5	16.0	20.2	23.5	35.2	45.7	49.2	60.4	64.7	62.1
GCC total	291.1	314.4	349.8	437.1	509.0	605.5	706.1	824.0	1,502.6	1,664.5	1,911.7	2,210.9

Source: The Arab Monetary Fund, National Accounts for Arab States, various issues.

Table 5.4: Principal Tree Crop Production, Selected Years (thousands of metric tons)

	1969–71	1975	1978	1979	1980
Dates (Tonnes)					
Bahrain	15	16	38	38	38
Kuwait	...	...	1	1	1
Oman	44	55	51	52	53
Qatar	n.a.	n.a.	n.a.	n.a.	3
Saudi Arabia	161	140	210	215	205
UAE	8	n.a.	39	40	51
GCC total	228	211	339	346	351
Lemons and Other Citrus (Tonnes)					
Bahrain	1	1	1	1	1
Kuwait	...	...	...	...	...
Oman	9	12	13	14	14
Qatar	...	...	...	...	0.33
Saudi Arabia	n.a.	6	29	29	29
UAE	19	n.a.	4	9	4
GCC total	n.a.	n.a.	47	53	48.33

Note: n.a. indicates 'not available'; ... indicates 'negligible'.
Source: FAO, *Production Yearbook*, various issues; Qatar, *Agricultural Statistics Yearbook*, 1980.

Table 5.5: Principal Cereal Crops, Area Planted, Yield and Production in Oman and Saudi Arabia, Selected Years

	1969-71	1975	1978	1979	1980
Oman					
Total cereals					
area	3	5	4	4	4
yield	1.2	0.97	1.47	1.52	1.52
production	4	5	6	6	6
Wheat					
area	1	3	2	2	2
yield	1.63	0.98	1.94	2.0	2.0
production	2	3	4	4	4
Barley	n.a.	n.a.	n.a.	n.a.	n.a.
Sorghum	n.a.	n.a.	n.a.	n.a.	n.a.
Saudi Arabia					
Total cereals					
area	357	345	406	424	424
yield	1.21	0.84	0.75	0.67	0.68
production	430	289	304	283	285
Wheat					
area	57	62	60	85	85
yield	1.78	2.13	2.0	1.8	1.8
production	101	132	120	150	150
Barley					
area	14	7	8	10	10
yield	0.9	2.4	1.9	1.6	1.6
production	13	17	15	16	16
Sorghum					
area	174	237	302	290	290
yield	1.06	0.54	0.5	0.35	0.35
production	185	128	152	100	100

Note: Area — '000 hectares; yield — tonnes/ha; production — '000 tonnes; n.a. indicates 'not available'.
Source: FAO, *Production Yearbook*, various issues.

Table 5.6: Selected Fruit and Vegetable Production Statistics in the GCC Countries, Selected Years

	1970	1975	1978	1979	1980
Tomatoes					
Bahrain					
yield	23.7	26.0	50.0	50.0	50.0
production	2.0	3.0	9.0	10.0	10.0
Kuwait					
yield	10.5	19.0	25.3	25.2	21.1
production	2.0	4.0	10.9	11.6	11.0
Qatar					
yield	n.a.	n.a.	n.a.	n.a.	17.6
production	n.a.	n.a.	n.a.	n.a.	6.0
Saudi Arabia					
yield	12.5	10.0	10.0	10.0	10.7
production	100.0	112.0	180.0	181.0	167.0
UAE					
yield	17.3	n.a.	34.0	34.0	46.0
production	9.0	n.a.	22.0	22.0	36.0
Sweet melons					
Bahrain					
yield	50.0	56.0	84.6	81.4	88.0
production	1.0	1.0	2.0	2.0	2.0
Kuwait					
yield	24.4	n.a.	22.5	67.5	n.a.
production	4.0	n.a.	4.5	2.7	5.0
Qatar					
yield	n.a.	n.a.	n.a.	n.a.	7.7
production	n.a.	n.a.	n.a.	n.a.	2.0
Saudi Arabia					
yield	32.7	n.a.	8.5	8.5	17.4
production	9.0	n.a.	12.0	12.0	16.0
UAE					
yield	27.0	n.a.	24.4	24.4	22.0
production	n.a.	n.a.	1.0	1.0	15.0
Water melons					
Bahrain					
yield	n.a.	n.a.	85.0	60.0	n.a.
production	n.a.	n.a.	1.0	1.0	1.0
Qatar					
yield	n.a.	n.a.	n.a.	n.a.	8.0
production	n.a.	n.a.	n.a.	n.a.	9.0
Saudi Arabia					
yield	34.2	n.a.	20.8	20.8	13.3
production	505.0	n.a.	260.0	260.0	140.0

Table 5.6: Selected Fruit and Vegetable Production Statistics in the GCC Countries, Selected Years (contd)

	1970	1975	1978	1979	1980
Water melons (contd)					
UAE					
yield	8.0	n.a.	35.7	35.7	31.0
production	1.0	n.a.	9.0	9.0	25.5
Total production ('000 MT)					
Bahrain					
Vegetables and melons	4	9	25	26	27
Fruit	18	19	42	42	42
Kuwait					
Vegetables and melons	15	16	31	32	33
Fruit	1	1	1	1	1
Oman					
Vegetables and melons	6	8	8	8	8
Fruit	65	79	82	83	85
Qatar					
Vegetables and melons	23	31	37	39	n.a.
Fruit	2	2	2	2	n.a.
Saudi Arabia					
Vegetables and melons	674	436	598	600	523
Fruit	259	434	438	446	520
UAE					
Vegetables and melons	24	n.a.	73	73	75
Fruit	28	n.a.	76	77	78
GCC					
Vegetables and melons	746	n.a.	772	778	666
Fruit	373	n.a.	641	651	726

Note: yield — MT/ha; production — '000 MT.
Source: FAO, *Production Yearbook*, various issues; Qatar, *Agricultural Statistics Yearbook*, 1980; UAE, *Statistical Yearbook*, 1980; Kuwait, *Statistical Yearbook*, 1980.

Table 5.7: Numbers of Livestock in the GCC Countries, 1976, 1977 and 1978 (thousands of animals)

	Year	Cows	Sheep	Goats	Camels
Bahrain	1976	5	3	13	1
	1977	5	3	13	1
	1978	5	3	13	1
Kuwait	1976	5	13	2	5
	1977	6	18	2	6
	1978	6	25	2	6
Oman	1976	134	76	194	6
	1977	135	77	197	6
	1978	136	78	201	6
Qatar	1976	10	38	42	9
	1977	10	39	39	9
	1978	6	42	49	9
Saudi Arabia	1976	321	2,243	1,577	107
	1977	360	2,300	1,600	107
	1978	400	2,600	1,700	108
UAE	1976	16	74	270	60
	1977	19	95	250	48
	1978	23	120	311	53
GCC total	1976	491	2,447	2,098	188
	1977	535	2,532	2,101	129
	1978	578	2,868	2,276	183

Source: ECWA, *Economic Indicators in the Arab World*, pp. 103-7.

Table 5.8: Agricultural and Total Development Expenditures by the GCC Countries, 1970-1980 (in current US$ million)

	Agriculture			Total			Agriculture as Per Cent of Total		
	1970-75	1975-80	1970-80	1970-75	1975-80	1970-80	1970-75	1975-80	1970-80
Bahrain	...	10	10	...	377	377	...	2.7	2.7
Kuwait	22	113	135	2,554	16,708	19,262	.9	.7	.7
Oman	24	165	189	1,305	3,926	5,231	1.8	4.2	3.6
Qatar	n.a.	n.a.	n.a.	3,472	4,584	8,020	n.a.	n.a.	n.a.
Saudi Arabia	444	1,332	1,776	7,280	90,520	97,800	6.1	1.5	1.8
UAE	164	191	355	3,472	4,584	8,020	4.7	4.2	4.4

Note: ... indicates 'negligible'; n.a. indicates 'not available'.
Source: The Arab League, The Arab Monetary Fund and The Arab Fund for Economic and Social Development, *The Unified Arab Economic Report*, 1981, pp. 293-4.

6 INDUSTRIALIZATION: PROSPECTS AND PROBLEMS

6.1 Introduction

In the Gulf, as in other developing countries, economic policy making involves substantial trade-offs between a variety of economic and social objectives not to mention political constraints and imperatives. A high rate of growth to increase the countries' productive capacities is a prerequisite of higher living standards, but it still leaves open the choice between consumption and capital formation, the present and future, the composition of sectoral outputs, the location of profits, the portfolio proportions of investment, etc. In addition, social considerations, an equitable distribution of income between rich and poor, between rural and city and between states of the region, may require a pattern of economic growth different from that which would maximize the rate of growth. Again, a choice has to be made between a more open or more self-reliant (or closed) economy, where openness is likely to have advantages in terms of efficiency and self-reliance in terms of security. All these trade-offs (policy choices) are relevant to planning for industrial development and must be faced seriously and expeditiously throughout the development process but more urgently in the initial stages.

A number of considerations have arisen recently that lead the policy makers of the Arab Gulf countries to give serious thought to the desirability of greater emphasis on manufacturing in general and non-oil manufacturing in particular. The likelihood that the oil sector will not continue to propel economic growth to the extent that it has done in the 1970s, because of both less favourable world prices and diminishing exportable volumes, poses the need for an alternative engine of growth, and manufacturing in the region is the most likely candidate, judging by all past development experience and given the factor endowments of the region.

In the GCC region, industrial development has lagged behind, partly because of the very abundance of oil and natural gas; and such industrial development as has occurred has been largely confined to the production of basic petrochemicals and non-durable consumer goods, with little if any production of basic metals, capital goods, high-technology products, etc.

There is indeed a strong case for the development of industry in the region. The conventional arguments for fostering industrial production in developing countries apply more strongly to the Gulf experience and endowments. There is first the comparative advantages that accrue from low-cost natural resources in abundant supply in the region. Secondly, manufacturing may generate the capacity to produce goods and machines embodying characteristics and specifications appropriate to their countries' factor proportions and tastes. Thirdly, there is always the prospect of externalities (spin-offs) from local production and industrial experience which can increase the efficiency of the entire economy. Fourthly, local productive capacity ensures a steady and reliable supply of critically needed products, in vulnerable periods at least. Fifthly, the region's manufacturing structure has the dubious distinction of being fragmented with short production runs and backward technical affinities. The promotion of manufacturing activity on a national and regional basis may allow the region to fill the enormous gaps and 'white spots' in its technical structure of production. This last point suggests that regional co-operation in industry may avail the region of the critical mass or scale needed to produce a diversified vector of goods economically, and thus the chance to redress the lopsided structures of their respective economies and manufacturing sectors.

For the industrial effort to be sustainable and self-perpetuating it must ultimately be cost efficient. Efficiency here encompasses both the static considerations of the marginal conditions of profitability and the dynamic considerations of viability and growth. The static conditions concentrate on comparative costs and efficiency scales. Industries in the region must be predicated on using local resources in abundant supplies. Naturally this is the main rationale for oil downstream activities discussed in Chapter 3. But oil is not the only natural resource and the vertical spread of economic uses of oil and gas are not the only economic uses these resources can be put too. There is a strong argument for another 'pole of growth' that utilizes the oil and gas as energy inputs and as feedstocks in iron and steel production, in refining alumina and smelting aluminium, in smelting copper, etc. Since the oil 'growth pole' has already been discussed, emphasis here will be placed on non-oil resource-based industrialization. In both cases, a long and steady stream of oil and gas is needed to fuel, feed and finance these poles. This calls for large and long-lasting reserves of oil and gas. Thus the mere availability of oil and gas is necessary but not sufficient. Equally important is the availability of large stocks of both that last for a long period. The GCC region, fortunately, has both in abundance.

Since labour in the region is in short supply, the industrial effort must economize on its use. Industrial activities outside oil should be ranked in terms of their direct and indirect labour requirements. Activities selected for further consideration should be those with the least total labour requirements, that have a large market domestically and are potentially capable of exporting their products. The emphasis on low labour intensities will eliminate a large subset of industries (textiles, consumer durables and even some capital goods). If labour-intensive industries are to be promoted on a non-economic basis, then the social costs of these industries must be well accounted for.

Water is another resource that is in short supply in the region. Again, industrial uses of water are many and substantial. A ranking of industries in terms of their direct and indirect uses of water must also be made. This brings another filtering criterion. Industries with large water and labour requirements should be eliminated from consideration. Industries selected for careful study should alternatively be those with low labour and water requirements. Added attributes should also be considered — the capital and energy intensities can also be ranked, the size of the domestic market and the relationship of domestic demand to the minimum efficient scale of production.

Although constructing such a master table to take into account all the considerations enumerated above is beyond the scope of this study, a number of cursory remarks about potential candidates can be made. For security reasons, a food production pole is essential. Of course, alternative security arrangements should be carefully weighed before domestic production is considered. Moreover, there are many compelling reasons for establishing a window on technological developments in the field of industry so as to maintain a relative technological status. Given the pivotal role machines play in facilitating and substantiating domestic production and technical advancement, a capital goods activity (however small) must be contemplated.

The details of the profile of domestic industrial production as it stands now and as it may be shaping up in the future are the subject of this chapter.

6.2 The Non-oil Manufacturing Structure in the GCC Region

It is pointless to address the issues of manufacturing within the context of individual GCC countries. The question of size is particularly relevant in the context of manufacturing not only in terms of market size but also in terms of resource pools and human skills. The focus here is of necessity on the region as a whole — on what can be achieved and on the constraints and difficulties. The order of discussion is organized in terms of poles of production.

6.2.1 Resource-based Industrialization

The strategy of basing industrial development on advanced stages of processing of natural resources is generally motivated by the desire to capture the high value-added component of such activities, to diversify production and exports, and to exploit such comparative advantages as may exist in the production of competitive commodities. Industrial processing may also contribute to several other development goals. Processing often entails high degrees of utilization of capital and energy, and this fits well with the resource endowments of the GCC region and with the desire to avoid increasing dependence on an already scarce supply of labour, especially skilled domestic labour. The diversification of exports is important because the markets in which finished products can be sold are more diversified geographically than those of crude oil and minerals. Hence, processing before export (or before re-export) might allow the GCC to capture some of the monopsony profits formerly absorbed by the heavily concentrated buyers of crude minerals and other raw materials. When the processing of raw materials is carried to the fabricating or manufacturing stages, it may also encourage local production of products not related to the original raw materials. In this way, forward and backward integration of the input-output structure of the economy might lessen the GCC's dependence on crude oil, and thus promote more generally the important objective of industrial diversification.

Access to expertise and technology is required for the design and implementation of investment programmes and for the operation of new plant and equipment. Furthermore, the market for the output must be assured. The collaboration of transnational corporations (TNCs) may be necessary, as these often hold strong monopoly control over technology and markets. But TNCs are not easily persuaded to relinquish their monopoly power, and they may be unwilling to share their knowledge with the developing countries. Moreover, the governments of industrial countries are anxious to protect the interests of their labour and capital employed in processing within their own territories, and in this their aims coincide with those of the TNCs. Such protection is commonly assured by imposing tariffs that rise with the degree of processing. In many cases, the effective tariff on value-added in processing is so high as to make profitable processing in developing countries very difficult.

Independence in industrial processing presupposes the development of national competence, in the form of skilled and knowledgeable individuals and appropriate institutions. The advantages of industrial development to the region tend to be much reduced in the absence of such competence.

The present world allocation of processing activities is, to some extent, also the outcome of the biased structure of transport costs. The savings in transport costs from reduced volume and weight of processed products is often not fully reflected in reduced transport charges imposed by shippers (especially conference shippers). The development in the GCC of shipping fleets may assist the development of processing activities in their area and in other developing countries, and the GCC is in an advantageous position in this respect.

The successful involvement of GCC countries in the processing of raw materials and semi-finished goods will depend on a number of interrelated factors. Three such factors will be decisive: imput availability; conditions of processing; and characteristics of output.

Input availability must be measured by comparative cost criteria. Raw materials and other complementary inputs are assessed in terms of their availability in sufficiently large quantities to make it possible to process them economically *in situ*. Whether they can be imported at advantageous prices, as an alternative to domestic supply, is another critical consideration (e.g. bauxite from Australia).

The conditions of processing are determined by the technologies used in the processing activities, and here there are three main considerations. The first pertains to the extent to which economies of scale facilitate or impede the locating of productive capacity in the region, because of the abundance or lack of abundance either of the raw material itself or of other complementary inputs. The second relates to the range of technological choice available within the industry, and possibly to the availability of processing systems that are particularly suited to the conditions of the region. The third has to do with the development of new technologies, or variants of existing ones, that may alter some of the circumstances militating against processing in the region.

The characteristics of output that are of special importance are those that determine the difficulties encountered in supplying end-products to foreign markets, including transport and storage problems, tariff and non-tariff barriers, and other difficulties associated with marketing and distribution.

A recent study has indicated that non-ferrous metals, industrial chemicals and petroleum refining are particularly low in labour intensity, when compared with other major production sectors.[8] A study of investment potential in developing countries showed all resource-based industries, except wood products, to have capital-labour ratios of three to ten times the average for all industries combined, and labour coefficients 33 to 80 per cent below the average.[9] Evidence such as this, although admittedly not conclusive, tends to suggest that this type of industrialization is suitable for the GCC region.

Some recent changes in technologies offer significant opportunities for GCC countries. An example is the direct reduction of iron ore into sponge iron, using natural gas instead of coke. The sponge iron can be reduced to steel in electric arc furnaces, on a very small scale, using inputs of scrap of various qualities. The heavy use of natural gas and electricity, and the limited minimum efficient scale of the direct reduction method, provide a formidable comparative advantage to the GCC region, which has abundant gas and thermal electricity potential, and can import ore cheaply.[10]

Smelting alumina into aluminium through electrolytic processes is again energy-intensive, as well as scale-efficient at low output levels, and capital-intensive. Moreover, the GCC region is advantageously located near Africa (particularly Guinea, which has over a third of the world reserves of bauxite). Similar advantages may be found in the smelting of copper and alloys (as is already the case in Oman).

Marketing opportunities in the Arab world, and in neighbouring Asian and African countries, are considerable. With a proper allocation of investment to activities in these regions, the GCC countries might bring into being a formidable procurement-distribution network, with benefit both to the GCC countries themselves and to the neighbouring regions.

Processing need not be confined to the region's own mineral deposits or other domestic resources. Raw materials or semi-finished goods may be imported for further domestic processing, and the products then used for home consumption or exported. This might be particularly desirable in the case of products that require intensive combinations of capital and energy and can be produced efficiently with relatively small-scale operations and limited water usage.

Traditionally, industrial development has been viewed from what may be called a 'horizonal perspective'. That is to say, attention has been focused on activities as they relate to products classified by major sectors of production or by categories of subsequent use. Analyses of manufacturing are usually carried out in terms of light and heavy industry, or in terms of consumer and capital goods. However, this approach makes it difficult to distinguish different activities within a given branch of industry. On the other hand, an examination of resource-based industrialization requires a 'vertical perspective': activities must be examined according to their stage of processing — in terms of primary, semi-finished and finished goods — rather than according to the characteristics of the final product. It is this perspective that will be used here.

Four basic industries have been selected for examination: iron and steel, aluminium, cement and copper. All of these activities are energy-intensive and could be based on natural gas which is in abundance in the region. Their water requirements are moderate and their production could be coupled with desalination plants.

6.2.1.1 Iron and Steel. Production of steel in the GCC region has not kept pace with consumption and the gap is expected to grow over the coming two decades. Whereas consumption exceeded 3.3 million tons, production in the region fell short of 0.7 million tons in 1978 (see Tables 6.1 and 6.2, pp. 122-3). Over the next two decades consumption is forecast to reach 17 million tons by 1995 whereas production is not expected to exceed 3 million tons (see Table 6.2).

Of the dozen or so combinations of processes for making iron and steel, the one which merits serious consideration by GCC member countries is the direct reduction of iron ore to make sponge iron followed by electric arc furnace treatment to produce finished steel. The direct reduction process to produce sponge iron and steel using natural gas has many advantages, including operating simplicity, lower capital costs compared to blast furnace operations, suitability for small-scale operation and non-reliance on imported scrap to make steel (Saudi ores could be beneficiated and used too).

The case for locating direct reduction plants in the GCC member countries rests mainly on the availability of low-cost energy — gas for the ore reduction process and relatively cheap power for an associated electric furnace steel plant. The price at which gas is made available to a new iron and steel venture is, therefore, a key to its profitability and future viability.

The question often arises as to whether production of iron and steel should be geared for domestic use or export. Given the large gap anticipated between production and consumption, this issue is not likely to be a significant one in the GCC region for some time to come.

The iron and steel industry does not operate in a vacuum and as such care must be exercised to co-ordinate the erection of capacity to produce these products in isolation of user and feeder industries. Balanced growth requires a complex strategy that co-ordinates all the phases involved. In particular the following industries need to be considered in a specific and systematic manner to meet the requirements of iron and steel plants in the region as well as consuming industries:

(1) castings and forgings;
(2) alloy steel plants to cater to the needs of engineering and manufacturing industries;
(3) graphite electrodes plants; and
(4) basic refractories plants.

Castings and forgings are the main raw materials needed by the engineering industry: hence their establishment should be encouraged and given high priority.

6.2.1.2 Aluminium. Aluminium production is energy- and capital-intensive. With low-cost energy, suitable geographical location, large investment funds and low environment sensitivity, the GCC region is considered well placed for production of this commodity. Low-cost electrical power will remain the key to economic production of this product and GCC member countries may, therefore, be able to compete by using their gas to generate the necessary power in gas turbines (see Table 6.3).

The primary aluminium industry, which comprises mainly smelters, is presently located in the GCC region in Bahrain and the UAE. The total capacity of both smelters in the region amounts to 255,000 tons (120,000 and 135,000 respectively).

Several other states are at the stage of planning new facilities, among which are:

Bahrain (expansion by)	45,000 tons
Qatar	150,000 tons
UAE	150,000 tons
Total	345,000 tons

In the event that all these plans are implemented, then total production capacity will approach 600,000 tons, which is in excess of the anticipated regional consumption level in 1990 (see Table 6.4). Therefore, the new plants have to be reviewed in terms of their access to world market and future price levels. Saudi Arabia had contemplated the construction of a smelter with 200,000 tons capacity. Recently it shelved the project preferring instead to become part-owner of Bahrain's ALBA.

Among fabricated products, rolling is considered a highly capital-intensive industry. Moreover, rolling products represent the largest portion of consumption of aluminium in the region. These products are normally: plates of 6.45 mm thickness and above, sheets of 0.15–6.32 mm thickness, foils of less than 0.15 mm thickness and sections. Other fabricated products are processed through castings, extruding, forging and drawings.

Several processing units related to extruding and drawings exist in the region, but only recently has a rolling mill with 40,000 ton capacity been agreed to by Gulf countries, as a joint-venture enterprise, to be erected in Bahrain. It was to be commissioned in 1983/4. The product will mainly meet local demand.

Apart from alumina, the other main materials required for production of aluminium are: cryolite, aluminium fluoride, electrodes or electrode materials (carbon and pitch) and petroleum coke, which materials represent more than 10 per cent of manufacturing costs (see Table 6.3). Therefore, studies should be carried out in order to find the viability of producing aluminium fluoride, cryolite and petroleum coke in the region in sufficient quantities to feed the existing and planned smelters.

6.2.1.3 Cement. This industry is perhaps the best suited for production in the region given that most of its raw material requirements are available locally. Moreover, gas is an ideal input in the production of clinker and cement, and gas is abundantly available. Nevertheless, until very recently the region imported large volumes of cement, but this was the result of a unique situation that is not expected to hold for long.

In the meantime, a large number of cement plants were constructed and now, except for Oman, every GCC state has at least one cement plant. By June 1982, the GCC *installed capacity* amounted to over 14 million tons and this is expected to rise to almost 24 million tons by the early 1990s (see Table 6.5). Should local demand moderate its growth, the region may be in a position to export its surplus to other neighbouring countries. Given the high transport costs of cement and excess capacity for production in neighbouring countries, it is not likely, however, that much in the way of exports can be anticipated from the region and as such capacity extensions should be undertaken with special care.

The choice of technology lies between a wet or dry process. For most GCC countries, the dry process should be chosen because of its lower water requirements as well as lower energy consumption.

6.2.1.4 Copper. The GCC region contains a large tonnage of poly metallic sulphides. The principal minerals are pyrite, chalcopyrite and sphalerite, which are, respectively, sources of iron, copper and zinc. Copper is actively mined at the Sohar region of Oman and is smelted into about 20,000 tons per year there; further processing of copper by electrolytic refining (20,000 tons per year) is scheduled to take place in the near future. Copper mining is potentially economic in Saudi Arabia too. For example, the Jebel Saeed deposit could be economically mined; it could produce from 1 to 1.3 million tons of ore per year, depending on the method of exploitation.[11]

A second example is the Nuqrah area deposit in Saudi Arabia which could be economically exploited by underground mining for a period of up to 12 years at a rate of up to 140,000 tons of ore per year. The deposit is five to seven times richer, in terms of ore value, than other copper deposits in most developing countries.

The current practice of copper mining, smelting and planned future refining in Oman has produced some experience in the domain of integrated copper production, which will be useful in planning any further development in the GCC region.

Whereas moving downstream and upstream in the oil and gas sector represents exploiting vertically the comparative advantage of the region, developing the mineral resources of the region, or further processing within the region of imported ores, represents exploiting horizontally the comparative advantage of the region in oil and gas, capital abundance and availability of raw materials. There are two added dimensions, of comparative advantage, to this horizontal extension — first, the potential use of local ores and secondly, utilizing complementary inputs usually produced as by-products of oil and chemical refining in the region. As examples of the first dimension it may be possible to mention the iron ores in Wadi Sawawin, Jebel Idsas and Wadi Fatima in Saudi Arabia. On the other hand, petroleum coke and caustic soda are illustrative examples of the second dimension.

The region's abundant energy supplies, its prime geographic location and its sparsely populated areas qualify it to embark on a serious evaluation of major extensions and expansions in the direction of further processing local and imported minerals.

6.2.2 Food and Agricultural Processing

Food security considerations, raising the embodied domestic value-added in local resources and encouraging the private sector to capture industrial opportunities provided by the larger GCC market, combine to make food and agricultural processing a potentially desirable activity in the region.

Processing opportunities for traditional agricultural products, however, are fairly limited. The bulk of domestic production serves local needs and is often produced, processed and consumed within a local market area. At this level, given the general lack of statistical information in the region, there is a considerable amount of uncounted activity. Data-gathering procedures do not operate well at this low a level, particularly if the exchange process has no records. Traditional cash crops, such as dates and citrus fruit, require little in the way of processing prior to sale, either domestically or on the export market. Thus, agricultural processing is based primarily on expanded production of vegetables, livestock and other non-traditional crops, or on the processing of imported raw materials, such as cereals and semi-finished products. Two types of processing structures can, therefore, be distinguished. The first involves a forward linkage from domestic production, and depends for its growth on the expansion of local agricultural output, either crops or livestock. The second is oriented towards transforming raw materials purchased abroad into final products.

Although the agricultural processing industry has several general characteristics that would appear to make it an unlikely sector for major growth within the region, on balance there are still several net benefits. In many cases, primary agricultural inputs to the industry are highly perishable. This eliminates the possibility of importing the raw material for local processing. For these types of processing facilities, such as meat packing and vegetable canning, any plants would have to rely on local supplies. In order to produce at prices competitive with processors in other areas of the world, the local industry would need heavy subsidies or low raw product prices. If, however, raw product prices are low to the farmer, so that the costs of production exceed the returns, the output will not be produced and the industry will be unable to operate. Thus, the implication is that, unless farmers can profitably produce the product at a low price, the processing industry will require subsidization or tariff barriers to inhibit foreign competition, at least initially until such time as the domestic industry is mature enough to compete on an equal footing with foreign suppliers.

A second consideration is that most processing activities tend to be weight-reducing: the product loses weight during processing. Thus, waste is produced and discarded. This typically leads to processing plants locating close to the primary sources of raw agricultural products in order to take full advantage of reduced shipping costs. For the GCC states, processing plants that rely on imported raw materials (such as oil seed crushing plants) will be high-cost producers unless a relatively high value can be assigned to their by-products, and these are many and often substantial. Otherwise, all of the large shipping bills associated with transporting the raw material will have to be assigned to the cost of producing the desired products.

Ultimately, the factor that most limits expansion of agricultural processing is the high cost of water in the region. Traditional agricultural processing plants use high volumes of water in virtually every stage of processing. Where the costs of water are high, processing of food products is a high-cost activity.

All of these influences place severe limitations on the growth potential of the food processing sector, particularly as an export-oriented industry. Opportunities for expansion of the sector will arise primarily through import substitution. In this vein, Oman has recently completed a feed mill in Muscat to formulate feeds for livestock. This will reduce the need to import such feeds. To the extent that local agricultural production can be increased and made available, a small but viable processing sector could develop. This is most likely in the poultry processing and vegetable processing sectors. Recent advances in recycling technology have reduced the water requirements for food processing plants. Using advanced technology, a net surplus of water can be produced.[12] Of course, the costs of this technology are high, but such plants may appear viable when account is taken of the price of importing the processed product. Similarly, industries based on imported raw materials may be viable operations within the region if their waste products are not excessive.

With the great increase in population in the region, considerable opportunity exists for the beverage industry. Since the production process consists of adding water to a concentrate prior to packaging, the transport costs of the raw materials are low. Furthermore, the final product sales price is relatively high and will more than cover the true value of the water.

To the extent that the region's population reaches a large enough size that processing plants can take advantage of economies of scale, there is opportunity for increased activity. However, the particular type of operation must be chosen carefully if costs are to be kept under control.

Expansion of fisheries is another activity with high potential in the region. A by-product of this activity is the large volume of by-catch. Approximately 8,700 tons of by-catch were available in the region in 1979. The growing production of broilers and eggs in the region provides a major opportunity for the utilization of this by-catch. Fish-meal can be used as a protein supplement in livestock feed. Expansion of existing feed milling facilities to produce feed using an indigenous product would reduce import costs and provide stimulus to local fisheries.

Current food and agricultural processing activity in the region is limited but data on this subject are particularly difficult to come by and to organize within a consistent framework. Consequently, the available material is presented on a country by country basis.

Bahrain possesses some food processing capacity but this would seem to be designed solely for domestic production and to rely primarily on imported raw materials. Data for Kuwait are available for 1977, in which year Kuwaiti authorities counted 346 bakeries,

16 confectionery producers, 14 producers of dairy products, seven animal feed producers and four bottling operations. Not surprisingly, bakeries dominate the sector in terms of numbers as they produce a highly perishable product which requires limited investment in capital or technology. Oman has developed fruit packing plants in Batinah and Salalah to take advantage of local production. In addition, there are new date storage facilities in Muscat. The country also has dairy facilities capable of producing 0.72 million litres of yoghurt and 1.5 million litres of milk products. Flour milling capacity has recently been expanded and a new animal feed plant has been constructed. There are also the usual bakeries and confectionery operations found in all communities.

Qatar possesses minor meat packing and dairy operations, in addition to bakeries and other local supply facilities. Saudi Arabia has the most extensive set of processing facilities in the region. Table 6.6 provides an indication of existing and planned operations; and also indicates a diverse set of activities, with significant expansion of capacity planned. Firms enumerated in this list are major producers, not small local neighbourhood establishments. No information on food processing activities was obtainable for the United Arab Emirates.

6.2.3 Capital Goods and High-technology Products

Aside from natural resource-based industries and the other growth poles mentioned above, there are many industries which are likely to be viable candidates for co-operation among GCC member countries. One such area is in the field of heavy engineering industries, which we refer to here as the capital goods pole. These industries are characterized by quite heavy capital investment, highly skilled manpower and a long maturation period. Another outstanding characteristic is the existence of a large number of links between their output and that of other productive activities (forward and backward linkages). The linkages of these industries lend themselves also to split production, whereby certain products produced in one country of the GCC can be used as inputs in other plants located in another GCC country.

A review of the structure of manufacturing output in the GCC member countries reveals that more complex and integrated activities in the engineering industries have only recently been given attention by some member states. The lack of financial resources, the limitation of the market at the country level, and the absence of a co-operative outlook among the member countries in the past may have inhibited the establishment of such industries.

With the exception of Saudi Arabia in the case of a limited number of industries, the GCC member countries individually will not be able to develop viable industries in this field because each member country, when operating alone, lacks one or more factors necessary for the development of viable and dynamic engineering industries. Thus it would appear that in the GCC countries collectively, the high level of demand for capital goods which has been created by their ambitious development programmes, can serve to encourage the rapid expansion of the capital goods industry within the region. As an indication of magnitude of demand, we note that the region's import of engineering products increased from US$1,392 million in 1973 to US$13,420 million in 1978.[13] During this period, the member countries spent an average of close to 14 per cent of their GDP on imports of such products. Present indications suggest that such a trend will continue for some time to come: the forecast for 1990 puts the amount for engineering products in the GCC region at US$22 billion. We therefore use most of the rest of this chapter to provide a preliminary exploration of certain capital goods industries which hold out some promise as possible candidates for expansion and for import replacement, and whose development would require co-operation among the GCC member countries.

Our intention is not to provide an exhaustive list of possibilities. Instead, we attempt to provide suggestions for the sorts of industries which would seem to build on the development of the region and, in turn, would encourage further development. The capital goods manufacturing industries which we discuss produce: (i) telecommunications equipment; (ii) electric power equipment; (iii) telephone and power cables; and (iv) machinery and equipment for the chemical and petrochemical industries.

6.2.3.1 Telecommunications Equipment. Telecommunications equipment includes a wide range of products extending from simple ones, such as telephones, to complex ones, such as broadcasting equipment, radar, microwave equipment, exchanges and telex machines. However, since the major telecommunication services in the GCC member countries are likely to remain concentrated in telephones and telex, they are the ones of concern to us here.

Many problems face the establishment of a telecommunications industry, including the acquisition of technology and the need to keep up with new products. At the same time, such an industry would stimulate the development of other strategic industries, notably the electronics industry. Furthermore, a major prerequisite to the success of this industry at the GCC level is close co-operation among the governments of the region, since the industry has strong links with the government telecommunication authorities.

The growth of demand for telephone and telex equipment has been projected for the GCC member countries in Tables 6.7, 6.8 and 6.9, and is summarized below:

Average Annual Demand in the GCC Member Countries

	1981-5	1986-90	1991-5
Telephone exchange lines	252,000	511,000	610,000
Telephone instruments	327,000	664,000	792,000
Telex exchange lines and telex machines	4,500	5,980	n.a.

The projections indicate that the market of the GCC region could now absorb about 250,000 telephone exchange lines per year, which is considered by international manufacturers as the current minimum viable scale of production. In 1980 the ECWA/UNIDO Industry Division concluded that a production level of 250,000 lines per year would be economically viable in the Economic Commission for Western Asia (ECWA) region. Fixed capital per plant was estimated at about US$85 million with working capital at about US$24 million.

The trend towards electronic digital exchanges will continue in telephone systems, and future developments will reflect mainly improvements in electronic intermediate devices. The manufacturing, however, covers both hardware and software production and should be pursued in collaboration with international manufacturers. A detailed feasibility study would be required.

The production of telephone instruments is simple compared to other telecommunication products, and does not require foreign technical assistance since most of the needed skills are available locally. The minimum annual scale of production for an integrated plant is about 200,000 instruments per year. Such a level of output could easily be absorbed in the region. Alternatively, assembly-type operations of smaller plant capacities (e.g. 50,000 instruments yearly) could be produced at the national or at the GCC level. The major justification for manufacturing telephone instruments in the GCC region is to take advantage of backward linkages by manufacturing the components too. However, it may be that a central activity could feed assembly-type operations at a country level. A comparative study might be conducted to compare the advantages of an integrated industry against the alternative of several assembly plants which are supplied from one central facility.

The market for telex machines, even on the GCC regional level, is not adequate to support a minimum economic scale of about 100,000 units assembled annually. Therefore, it should not be considered for further study at the present time.

6.2.3.2 Electric Power Equipment. The electric power programmes in the GCC member countries are massive and dynamic. Table 6.10 and 6.11 present the projected yearly additions of generation and transformer capacities in each of the member countries; the projections are summarized below for the region as a whole:

Average Annual Increase in Capacity in the GCC Member Countries

	1981-5	1986-90	1991-5
Installed generation capacity (MW)	1,451	2,610	2,105
Transformer capacity (MVA)	7,637	14,165	11,349

The generation programmes are predominantly thermal, steam and, to a lesser extent, gas. Nuclear power generation is not likely to become significant in the next two decades.

Equipment for electric power generation and distribution includes generators, turbines, transformers, motors, switchgears and capacitors. Turbines and generators are the heart of the generation programmes, and are also the most costly items. (Their combined cost amounts to about one-third of the total cost of the thermal power plants.) Transformers are also relatively expensive.

The projected demand for turbines and generators in the GCC region over the next decade or so is at about the output level associated with a 2000 MW manufacturing plant. That is about the minimum plant size which international manufacturers consider to be economically viable. None of the countries alone could absorb the output of a plant this size, even by the year 2000.

If production in the GCC is to take place, it would be most important to standardize the sizes of the generating units. Available evidence suggests that most plans call for steam units with capacities of 100, 150 and 300 MW, and gas turbines units of 50 MW are the most likely. In addition, steam turbines will be required for electric power generation in industry, but in smaller sizes.

The required production machinery and equipment is quite expensive and complex and the technology, though stable, is also complex. It should be mentioned that the machinery required for producing 300 MW units, for example, is also capable of producing larger units up to 500 MW units, say. Moreover, the same machinery can produce industrial turbines and generators as well as a variety of motors.

Probably the first step in this area should involve the manufacture of generators, and a feasibility study could be done.

Referring to the projected demand for transformers, it is seen that the regional market for small distribution transformers (0.25-1.25 MVA) and medium power transformers (6-40 MVA) could, even now, absorb the output of a 3000 MVA manufacturing plant, the size of plant generally considered by international manufacturers to be the minimum that is economically viable. As for large power transformers (72-200 MVA), the projected markets could absorb the output of a manufacturing plant of the minimum viable size of about 600 MVA starting in about 1986.

From a manufacturing point of view, various sizes of transformers could be produced in the same factory. Furthermore, transformers for instruments, industrial uses and other purposes could also be produced in the same factory. However, the efficient production of transformers in the region requires extensive preparations relating to the standardization of transformer sizes and the evaluation of existing transformer factories, among other things.

In Saudi Arabia, several companies were licensed in 1979 and 1980 to manufacture transformers. Their total capacities exceed 300 MVA. Details on size are not known, but it is most likely that all these factories intend to produce small distribution transformers. Co-ordination of the output of these plants, as well as of those intended to be erected in the UAE, is essential at this stage. A study should be carried out to evaluate the existing licensed facilities in order to formulate a general plan for the manufacture of transformers in the GCC region.

6.2.3.3 Telephone and Power Cables. In the expansion of the telephone system, the direct exchange lines, as well as the trunk networks, will probably be based on cables. The technology for manufacturing telephone cables is not complex, nor are the skills required at the shop floor level of a very high order. However, the raw materials and the production management skills must be of high quality. We note that some raw materials, such as polyethylene and polyvinyl chloride, will soon be available from the petrochemical complexes in the region. The facilities for manufacturing telephone cables which now exist in the member countries are in plants which have a wide product mix, including the production of power cables. Thus the manufacture of telephone cables cannot be considered in isolation from that of power cables.

Many suppliers of power cables have been established in connection with massive electrification programmes undertaken in the GCC member countries. Bahrain has an operating cable manufacturing plant with a capacity of 14,000 tons per year, producing mainly all-aluminium conductors and steel-enforced aluminium cable (ACSR). The Gulf Cable and Electrical Industries plant in Kuwait produces low tension (LT) cables of various sizes, in addition to telephone cable. The production is projected to reach 20,000 tons/year. Plans are under way to manufacture cable joints as well as medium voltage cables. Six companies were licensed in Saudi Arabia in 1980 to produce cables and wires. Their total licensed capacity exceeds 90,000 tons/year. Details as to the type and voltage are not available, but over 30,000 tons/year were intended to be insulated overhead aluminium cables. It is also reported that a small plant producing aluminium conductors is in operation in the UAE.

Although detailed information about the operating plants in the region is not available, it is likely that they all produce low tension cables, telephone wires and bare aluminium conductors. The demand for high tension cables is still, by and large, met through imports. There would appear to be considerable scope for co-ordination between the existing and planned production units in order to avoid duplication and to achieve specialization of each plant in certain products, and thereby to improve performance and facilitate the creation of plants for manufacturing medium and high tension cables as well as telephone cables. A detailed study should be conducted for this purpose.

6.2.3.4 Machinery and Equipment for the Chemical and Petrochemical Industries. Included in this group are process vessels, storage tanks, bins, heat exchangers, furnaces and kilns, filters, pumps, compressors and boilers. The demand for machinery and equipment of this sort depends primarily on the development of such end-use industries and processes as gas extraction and processing oil refining, fertilizer production, petrochemical production and water desalination. According to surveys reported in the *Hydro-Carbon Processing Journal*, in 1978, the cost of fabricated static equipment (such as vessels, furnaces, heat exchangers) represented 37 per cent of the total cost of battery limit of the sorts of plants mentioned above, except those for water desalination.

The massive hydrocarbon and desalination projects planned by the member countries over the next two decades ensure a market for such equipment which is certainly large enough to justify minimum plant size. However, one of the major problems in developing such an industry is the tendency on the part of the industrial development authorities to adopt the short-cut approach of 'turn-key' projects, which makes it difficult for local manufacturers to break into the domestic market. Many companies in the GCC member countries have faced this problem. The case of the Kuwait Industrial Refinery Maintenance and Engineering Company (KREMENCO) is a good example. Although the company is well established and possesses all the necessary facilities to manufacture heat exchangers, it has been unable to market such equipment. In consequence, the activities of the company have been confined mainly to repair and maintenance work, as well as some fabrication of metal products.

The Arabian Ship Repair Yard in Bahrain is now underutilized. For this reason, the Kuwait Desalination Plants Fabrication Company intends to enter into a joint venture with it to fabricate the metal products (such as heat exchangers) required for desalination plants. The Kuwait Desalination Plants Fabrication Company, which has been licensed since 1979, is considering the fabrication and manufacture of static parts for desalination plants. At this time, it is licensed to sell only in Kuwait.

In Saudi Arabia, many companies have been licensed to produce pipes and heat exchangers; their licensed capacities to produce pipes exceed 120,000 tons/year.

In view of the difficulties faced in marketing heat exchangers, it is recommended that a detailed examination be carried out of the conditions which would ensure the satisfactory development of industries to produce fabrication facilities. Special attention might be given to the Kuwait Desalination Plants Fabrication Company, and to the steps which would allow it to expand from the national to the GCC level.

6.3 Conclusion

Pointing to the potentialities of a selective approach to industrialization does not mean that we can simply forget about balance in the development process and functions. We cannot ignore either the technical input-output relations of production or the demand patterns of the ultimate consumers.

The process of and the planning for industrialization must be holistic and comprehensive. Alternatives must be weighed against one another and complexes and networks must be considered simultaneously and together.

The end of the oil era is in sight and the gestation period of development of the region could extend beyond the life span of hydrocarbon resources of the region. Selectivity, carefulness and co-operation are key functions for the success of the regional development effort.

Table 6.1: Consumption of Iron and Steel, 1978 ('000 tons per year)

	Plates and Sheets	Tubes and Pipes	Rods, Bars and Wires	Angles and Shapes	Others	Total
Bahrain	11	11	25	4	1	52
Kuwait	29	161	196	15	16	417
Oman	5	8	37	–	–	50
Qatar	9	16	43	6	11	85
Saudi Arabia	183	382	710	111	739	2,125
UAE	68	132	240	54	150	644
GCC total	305	710	1,251	190	917	3,373

Source: Yearly Bulletins of Foreign Trade Statistics of the GCC countries.

Table 6.2: Arab Gulf: Profile of Steel Industry, 1982 ('000 tons)

	Existing Capacity					In Pipeline	
	Integrated Plants Steel Output	Rolling Mills/ Products	Foundries/ Forging Plants Finished Products	Tube Mills (spiral)	Special Steel Rolled Products	Under Construction	Under Study Planning
Bahrain	—	—	—	—	—	Pelletizing plant (2,000)	Special steel (1,000)
Kuwait	—	—	—	44	—	—	Pre-feasibility study: 140–240: sponge iron
Oman	—	—	—	—	—	Integrated plant (150) Phase I	Expansion of Phase I by 330 as Phase II
Qatar	400	—	—	—	450	—	Doubling capacity
Saudi Arabia	—	45	—	—	—	Integrated plant (800)	—
UAE	—	32	—	—	—	—	Integrated plant (500–800)

Source: Gulf Organization for Industrial Consultancy.

Table 6.3: Comparison of Investment and Operating Costs Per Tonne of Aluminium, 1980 Prices (all figures in US$)

	USA	Canada	Australia	Brazil	Arabian Gulf
Alumina	429.00	429.00	399.75	399.75	448.50
Petroleum coke	58.50	58.50	62.40	72.15	64.00
Pitch	24.50	24.50	25.50	29.50	26.50
Electricity	411.80	210.60	241.40	252.00	101.40
Maintenance	50.00	54.00	52.00	69.00	87.50
Direct labour, general administrative and other costs	185.00	170.00	173.00	173.00	190.00
Delivery	33.06	44.09	66.12	55.12	66.12
Total operating costs	1,191.86	990.69	1,020.17	1,052.52	984.02
Total cost per tonne including capital recovery @ 5%	1,393	1,208	1,229	1,329	1,245
10%	1,486	1,308	1,325	1,457	1,366
15%	1,592	1,422	1,435	1,603	1,503
Average market price 1980 ($) per tonne	1,715	1,715	1,715	1,715	1,715
Electricity costs As % of total operating cost	34.5	21.3	23.7	23.9	10.3

Source: Gulf Organization of Industrial Consultancy.

Table 6.4: Consumption of Semi-finished and Finished Aluminium (in tons)

	1979	Forecasted 1990
Bahrain	2,735	5,810
Kuwait	10,760	24,870
Oman	2,565	7,600
Qatar	2,590	6,045
Saudi Arabia	47,150	130,870
UAE	11,750	29,740
GCC total	77,550	204,935

Source: Yearly Bulletins of Foreign Trade Statistics of the Countries of the GCC.

Table 6.5: Arab Gulf: Cement Production Capabilities ('000 tonnes)

	On Stream (1982)	Planned Under construction	Planned Under study/ construction
Bahrain	400	–	1,000
Kuwait	1,425	–	–
Qatar	330	–	1,000
Oman	–	625	200
Saudi Arabia	8,955	2,675	2,600
UAE	3,100	2,670	–

Source: Gulf Organization for Industrial Consultancy.

Table 6.6: Selected Licensed Food Processing Factories in Saudi Arabia, 1980

Type	Principal Products	Number of Operational Factories	Number Under Construction	Number Planned	Total
Livestock	Eggs, broilers	5	4	2	11
Meat packing	Poultry, beef, sheep and camels — cutting, wrapping, processing, freezing	6	9	3	18
Dairy	Raw milk, pasteurized milk, butter, ice cream, cheese, yoghurt	29	21	28	78
Fruit and vegetable processing	Fruit juices from concentrate, water bottling, tomato paste, juice and ketchup, jams, date packing, canned vegetables	14	6	10	30
Bakeries	Bread, rolls, macaroni and processed rice	46	26	21	93
Sugar refineries	Sugar	0	0	2	2
Cocoa, chocolate and confectioneries	Cocoa, chocolate and confectionery items	8	2	3	13
Ice	Ice and cold storage	37	12	14	63
Beverages	Carbonated beverages, fruit juices, mineral and carbonated water	19	3	14	36
Feed mills	Poultry and cattle feed	8	0	3	11

Source: Saudi Arabia, Ministry of Industry and Electricity, *A List of Licensed Factories and Industrial Licences Issued Under the National Industries Protection and Encouragement Law and Foreign Investment Law up to 1980*.

Table 6.7: Projected Average Annual Demand for Telephone Exchange Lines

	1981-5	1986-90	1991-5
Bahrain	7,000	10,000	16,000
Kuwait	48,000	108,000	160,000
Oman	11,000	22,000	42,000
Qatar	6,000	11,000	12,000
Saudi Arabia	140,000	300,000	300,000
UAE	40,000	60,000	80,000
GCC total	252,000	511,000	610,000

Note: These projections are comparable to the projections of ITU Inc. in *Middle East and Mediterranean Telecommunication Project, Master Plan, Final Report*, December 1978.
Source: *The Viability of Establishing a Regional Telecommunication Industry in the ECWA Region*, Joint ECWA/UNIDO Industry Division, 1980.

Table 6.8: Projected Average Annual Demand for Telephone Instruments

	1981-5	1986-90	1991-5
Bahrain	9,000	13,000	21,000
Kuwait	62,000	140,000	208,000
Oman	14,000	29,000	54,000
Qatar	8,000	14,000	15,000
Saudi Arabia	182,000	390,000	390,000
UAE	52,000	78,000	104,000
GCC total	327,000	664,000	792,000

Source: See Table 6.7.

128 *Industrialization: Prospects and Problems*

Table 6.9: Projected Average Annual Demand for Telex Exchange Lines and Telex Machines

	1981-5	1986-90	1991-5
Bahrain	400	680	n.a.
Kuwait	600	1,000	n.a.
Oman	560	780	n.a.
Qatar	340	420	n.a.
Saudi Arabia	1,600	2,400	n.a.
UAE	1,000	700	n.a.
GCC total	4,500	5,980	n.a.

Source: *Middle East and Mediterranean Telecommunication Project, Master Plan, Final Report*, ITU, December 1978.

Table 6.10: Projected Average Annual Demand for Installed Generation Capacity (in megawatts)

	1981-5	1986-90	1991-5
Bahrain	76	127	83
Kuwait	325	499	477
Oman	14	14	16
Qatar	62	62	44
Saudi Arabia	750	1,548	1,164
UAE	224	360	321
GCC total	1,451	2,610	2,105

Source: See Table 6.7.

Table 6.11: Projected Average Annual Demand for Transformers

	1981-5	1986-90	1991-5
A. Small Distribution Transformers (0.25-1.25 MVA)			
Bahrain	91	156	100
Kuwait	390	600	573
Oman	17	17	20
Qatar	75	76	52
Saudi Arabia	891	1,860	1,398
UAE	269	450	384
GCC total	1,733	3,159	2,527
B. Medium Power Transformers (6-40 MVA)			
Bahrain	182	302	199
Kuwait	655	1,008	955
Oman	33	33	40
Qatar	150	149	104
Saudi Arabia	1,184	2,512	1,865
UAE	449	709	643
GCC total	2,653	4,713	3,806
C. Large Power Transformers (75-200 MVA)			
Bahrain	—	—	—
Kuwait	515	792	766
Oman	—	—	—
Qatar	—	—	—
Saudi Arabia	2,378	4,931	3,733
UAE	358	571	517
GCC total	3,251	6,294	5,016

Source: See Table 6.7.

7 CONCLUSION

The member countries of the GCC are determined to make effective use of their current oil-based wealth so that future generations will also be able to enjoy high standards of living. It is difficult to see how this goal can be realized by chance. Instead, the careful drawing up and implementation of realistic plans are essential for the diversification of the GCC economy before oil supplies are depleted.

The GCC region is the world's largest oil producer and has the world's largest pool of proven crude reserves. It also has very large reserves of natural gas. Moving 'downstream' to integrate the various phases and sequences of production — refining, processing, shipping, marketing — is an obvious and very promising way to encourage industrialization. Upgrading the domestic value-added component of oil- and gas-based products would provide a productive vent for the large financial surpluses of the region, diversify the markets for their products, and vertically integrate their production sequences without violating the static economic imperatives of comparative advantage.

Extensions of production into gas liquefaction do not seem to offer much advantages as the transformation and delivery costs are high, and there is a risk of the region providing, at low costs, feedstocks to foreign competitive ventures of their contemplated domestic facilities.

The production of petrochemicals is, however, more promising but care must be exercised initially and the region is better advised to concentrate on mature products for which the highest fraction of cost is the value of raw materials and the vulnerability to technological obsolescence is least. At the same time sufficient funds should continue to be committed to build large regional petrochemical complexes to substantiate the 'credible threat strategy' of persuading competitors of the seriousness of GCC producers' intention to penetrate world markets in these products. This necessitates pooling resources, harmonizing production strategies and allocating production facilities as part of a comprehensive and uniform regional plan. Size is a critical consideration in petrochemical production, and duplication of investment within the region is harmful to all and could undermine the region's favourable bargaining position based on long-term availability and low-cost raw materials.

132 Conclusion

The choice of joint ventures with TNCs should be complemented, perhaps on a larger and broader scale, with joint processing complexes in the rest of the Arab countries and in other parts of the Third World.

While the range of petrochemical products should probably be narrow for the short term, this need not be the case for the long term; in the foreseeable future olefins should be interfaced with the production of aromatics to broaden the production base and to move down the production chain from basic to intermediate and final petrochemical products, particularly to synthetic fibres.

Last but not least in this area, it is important to co-ordinate trade policy with planned investments in the petrochemical industry; it is difficult to see Western markets closed to Arab Gulf petrochemical products while Gulf markets remain open to all sorts of Western products.

The GCC region is not particularly rich in known mineral deposits, but a number of deposits have been identified which warrant commercial exploitation including iron ore, bauxite, copper and gold. The region, however, is particularly well suited for metal refining and smelting, inasmuch as these activities involve heavy commitments of capital, high level of energy use, oil and gas by-products in processing, and also because the region is well located between the sources of ores and the markets for final products and given that the region is generally sparsely populated and, therefore, is less sensitive environmentally to such processing.

The GCC minerals and mines ministries should complete a comprehensive mineral survey of the region; Saudi Arabia has such a survey, and the UAE has surveyed some of its territory; such surveys need to be completed, updated and harmonized for the entire region, and their results channelled to a regional registry.

There is a definite and urgent need to consider the forward and backward linkages involved in the processing of minerals to strengthen the horizontal integration of the structure of production of the region. For example, ALBA in Bahrain and DOBAL in Dubai could obtain some of their bauxite from deposits in Saudi Arabia, Saudi iron ores could be beneficiated with imported ores (particularly from Mauritania) to be used to produce iron in the region using the direct reduction method.

The GCC governments or even GCC private concerns are advised to consider opportunities of investment in mines and natural resources in the Third World countries, particularly Arab and Islamic countries.

These investments could assure a committed and steady supply of raw materials for processing in the GCC region.

The region is particularly rich in non-metallic minerals. There are large reserves of gypsum, limestone and clay which could satisfy most of the raw material requirements of the regional cement industry. Moreover, the region possesses extensive reserves of sand and gravel to support the massive regional construction activity, besides high quality quartz sand available particularly in Saudi Arabia could support a sizeable glass manufacturing activity.

The calcareous phosphate deposits in Saudi Arabia could also form the basis for regional phosphate, phosphoric acid and fertilizer industries. The required sulphuric acid, in turn, could be provided from the sulphur which is a by-product associated with oil and natural gas production. Furthermore, the high salinity sea water, and the extensive rock salt deposits in the region, could be used to support an inorganic chemical complex to produce caustic soda (useful in alumina refining), magnesia and magnesium.

The quest for diversification is partly dictated by considerations of balanced growth, substitution for oil and integrated development. It is also partly predicated on a desire to enhance the region's security. In this regard food security is critical and plays a major role in motivating agricultural activity. The efficient expansion of the regional agricultural sector requires a number of prerequisites:

(1) skilled manpower to operate, maintain and manage production and implements;
(2) the adoption of highly sophisticated irrigation practices and livestock management, reflecting the scarcity of water and of arable land in the region;
(3) the promotion of research to improve yields of native forages for use as livestock feed and to increase productivity of native livestock herds; research should also be directed towards adapting foreign crops and livestock to the high heat and salinity conditions that prevail in the area;
(4) the provision of extension services to inform farmers of new technologies, new crop varieties, and to inform agricultural researchers of the problems faced by farmers;
(5) the proper pricing of agricultural inputs, particularly water, and the stabilization of markets for farm outputs;
(6) measures to make better use of existing food supplies; a notable possibility is the use of animals sacrificed during holy Haj.

Food security calls for domestic production of at least suitable crops and the raising of suitable livestock. Many other crops and livestock are best acquired from the rest of the Arab world. The governments of the region, however, must set in place a procurement policy that takes advantage of favourable world market situations and maintains a diversified and a large food storage capacity.

The waters surrounding the region are rich in fish resources. A specialized agency might be empowered to manage the shared fish resources in the Gulf, Arabian Sea and the Gulf of Oman. Proper management would involve the development of an environmental policy to ensure a steady replacement of the marine life and the assurance of a clean water environment.

Water in the region is scarce and is primarily a shared resource, based on the common aquifers underlying the peninsula. The use of this resource is, therefore, a question of mutual concern. Joint planning of use and development is inescapable and should be practised urgently and carefully. Of special importance is setting the price of water at a level which reflects its scarcity value; such a policy would discourage reckless uses of this most valuable and scarce resource.

The GCC region should be viewed as a primary ring of production within the wider Arab ring. The Arab world is the economic, security and cultural depth of the region and every effort to harmonize and expand forward and backward links to it are bound to be mutually beneficial and vital. There is no contradiction between the GCC and the Arab region as a whole; both are complementary in much the same way as the Benelux countries' economic association reinforced and enriched the wider European Economic Community (EEC).

Within the regional ring, efforts need to be directed towards intensifying the mobility of goods, capital and people. The multiplicity of different local currencies complicates trade and investment within the region. A single currency area might foster the expected increase in intra-regional trade and investment. A GCC Payments Union should be considered as a means to strengthen the institutional mechanisms accommodating regional economic interactions.

Improving the transportation networks connecting the region is also required. Special care must be directed towards efficient and inexpensive modes particularly those exploiting the water bodies surrounding the member states. Synchronization of expansions with the expected increases in intra-regional trade volumes is imperative; it is totally wasteful to build large highways and airports beyond the expected demand levels.

Conclusion 135

The Andean countries' economic association provides an excellent example for the GCC to follow particularly in the field of Andean projects. The GCC has already set in place mechanisms and institutions to undertake and finance Gulf projects, but more is needed. The Gulf Investments Corporations need to be further expanded and developed.

The GCC is at a very fortunate historical juncture; an adequate infrastructure is now set in place, there is no need to forgo consumption to expand investment, there are no shortages of foreign exchange, and the constraint implied by the limited economic size of individual countries is being overcome through regional co-operation. The fact that water, skills and industrial experience are in short supply will present serious challenges to the collective will of the region in its attempts to shorten its development gestation period before its hydrocarbon resources are depleted. However, with careful planning, dedicated co-operation and time, these constraints which now limit growth can be relaxed, the opportunities for progress are real and must be realized.

The oil-based wealth of the region has not been without negative consequences and social costs. These costs have hitherto been underestimated. The collective will of the region must be exercised to overcome some of these ills and problems. A short list of these 'ills' include: consumerism, conspicuous investments and the development of rentier economies.

The sudden increase in oil revenues brought in its wake massive increases in consumption of all sorts of commodities and services. Consumption is the ultimate end of production, and the terms in which economic welfare is measured, but in the GCC consumption has outstripped the wasteful bounds of Veblen. The high consumption is divorced from local production and sometimes even from the engagement in productive work. The region has the dubious distinction of the highest *per capita* imports in the world.

This 'high consumption' without 'high production' has a number of negative consequences. First, it is extremely difficult to reverse these consumption trends and expectations once they are entrenched. Second, they are already a burden on the current accounts of most countries in the region. Third, social worth and status are now more sensitive to the size of one's villa and the number of cars one owns than the more sober traditional norms of personal and family conduct.

Large investments have been made in the region. A good number of these are productive and viable and in due course they will prove their worth. Equally true is the fact that many investments in the region are conspicuous 'white elephants' — large airports with capacities for multiples of the current populations. Some international airports are only walking distances from one another. This calls for a thorough review of investment projects, particularly through feasibility studies that cost these projects realistically and fully in terms of scarcity prices and social opportunity costs.

The extent and dangers of waste of wealth acquire serious dimensions when they involve human talent and initiative, and when this waste becomes ingrained in the operating systems of the economy. There are already significant warning signals that this might have already been taking place in the GCC countries. The divorce of income from production, the separation of consumption from local production and the relegation of the operations of the economy to expatriate labour are all symptomatic of the 'rentier economy'.

The diversification of the productive base of the GCC economies, which is vital, can only come with a sustained effort to industrialize and to upgrade the skills of the indigenous labour force. Some narrow strata of these societies can afford to live as 'coupon clippers' off the social dividends of oil, but the large majority of society cannot do this for any extended period of time with impunity. Work discipline, high productivity and rewarding production are economic imperatives for any advancing society.

The positive consequences of oil are still much higher than the negative consequences. But the positive benefits could still be augmented and the negative ones reduced and eliminated.

NOTES

1. The Gulf region and the Gulf Co-operation Council (GCC) region will be used interchangeably. Both comprise the six Gulf countries — Saudi Arabia, Kuwait, Bahrain, Qatar, UAE and Oman.

2. Former World Bank President Robert McNamara once suggested a depletion rate of 50 per cent of GDP.

3. Two points should be noted here: first, Oman and Bahrain are not members of OPEC; secondly, the neutral zone-share is not included in the GCC total. The latter share would increase output by 500 thousand barrels per day.

4. These figures are taken from Abdelaziz Alwattari, *Oil Downstream: Opportunities, Limitations and Policies*, Kuwait, OAPEC, 1980, p. 31 (Table 8).

5. The following equations were estimated for the GCC region

$$L_n\, NOGDP_t = a + bt + cL_n\, OGDP_t$$

where $L_n\, NOGDP$ is the natural logarithm of non-oil GDP
and $L_n\, OGDP$ is the natural logarithm of oil GDP.

The estimated values of the coefficients in the GCC region were as follows

$$L_n\, NOGDP_t = 5.62 + .203t + .247\, L_n\, OGDP_t$$
$$(2.22)$$
$$R^2 = .98$$
$$D.W. = 1.04$$

The figure in parenthesis is the t-statistic of the coefficient C.

6. This chapter is primarily based on UNIDO, *The Resource Base For Industrialization in the Gulf Cooperation Council Countries: A Framework for Cooperation*, mimeo, 1983, Chapter Five.

7. Robert Rowat, *Trained Manpower For the Agricultural Sector*, mimeo, FAO Mission to Saudi Arabia, Rome, December 1980, p. 37.

8. See J. Stern, 'The Employment Impact of Industrial Projects: A Preliminary Report', Discussion Paper No. 14, Harvard Institute for International Development, April 1976.

9. See M. Roemer, G. Tidrick and D. Williams, 'The Range of Strategic Industries in Tanzanian Industry', *Journal of Development Economics*, vol. 3, no. 3, October 1976, pp. 186-97.

10. See T.R. Stauffer, *Energy-Intensive Industrialization in the Arabian Gulf: A New Ruhr without Water?*, Harvard University, Centre For Middle East Studies, 1975.

11. United Nations Economic Commission for Western Asia, *Preliminary Survey Report on the Situation Pertaining to the Development of Mineral Resources in the Countries of the ECWA Region*, May 1977.

12. Reg Gallop, A.W. Hydamaka and A.P. Stephen, 'Conservation and Reuse of Water in the Food Industries per Multi, Integrated "Closed Loop" Process', in

Proceedings of the Water Reuse Symposium II, 1979, p. 847.

13. *Bulletin of Statistics on World Trade in Engineering Products*, UN/ECE, 1978.

INDEX

References to Tables are indicated by (T) following the page number.

Abu Dhabi National Oil Company 50
adhesives 48
agreements, Gulf States' 4-6
agriculture 75
    arable land: and crops, by country 87-9 (T); scarcity of 20-1, 76
    changes in 81, 133
    climatic constraints on 76
    costs 72, 82
    development expenditure on 82, 99 (T)
    employment in 77, 84, 90-2 (T); skilled 77-8, 82
    extension work in 84, 133
    food imports and 72, 82
    government intervention in 81
    inputs, pricing of 133
    investments in 80
    mechanization of, constraints on 77-8, 82-3
    production 77, 79; increase in 78, 82, by country 93 (T), by crop and country 94-7 (T)
    proportion of GDP from 9-10
    regional co-operation in 83-5
    research 83-4, 133
    sector linkages 75, 86 (Fig. 5.1)
    self-sufficiency possibility 81-2
    subsidies for 80
    yields 79, 133
    see also farm holdings, farm management, food processing, irrigation, livestock
aid for development, common policy on 5
airports 136
Al-Jubal Petrochemical Company 67
alumina
    refining 102
    smelting 106, 109
aluminium
    consumption by country 125 (T)
    production 108-9; international comparison of investment and operating costs 124 (T); materials 109, 124 (T)
ammonia 48, 50-1, 65-6 (T)
Andes economic association model 135
animal sacrifices for food 133
Arab Centre of Studies on Arid Zones and Dry Lands 83
Arab Company for the Development of Animal Wealth 83
Arab Economic Unity Council 83
Arab Organization for Agricultural Development (AOAD) 83
Arab World
    co-operation within 134
    joint processing complexes in 52, 132
    natural resources 132-3
Arabian Ship Repair Yard (Bahrain) 120
arable lands 87-9 (T)
    scarcity 30-1, 76
aromatics 48, 52, 132
assets abroad 15, 43

Bahrain 9, 11-13, 113, 119-20, 132
balance of payments, by country 36 (T)
balance of trade 11, 15
basic refractories plants 108
bauxite 105, 106, 132
benzene 48, 67 (T)
beverage industry 113, 126
butadiene 48
butane 48

camels 82, 98
capital cost differentials, Gulf and Europe 47
capital goods
    high technology 115-21
    imports of 115
    production 103, 114-15
castings and forgings 108

139

140 *Index*

catalytic reforming 48
caustic soda 111, 133
cement production 110, 125 (T), 133
cereals 79, 95 (T)
chemicals
    industrial 106
    industries 133; machinery and equipment for 120-1
commodity trade 13-15, 32 (T)
common market area 4
consumption, high 135
co-operation, regional
    for development 1-6, 52, 135
    in agriculture 83-5
    in industry 102, 115, 119
    in petrochemicals 131-2
    in telecommunications 116
    in use of common natural resources 72
copper 21, 110
    smelting 102, 106, 110
cost-efficiency 102
costs
    agricultural 72, 82
    see also capital cost, petro-chemicals projects
cows 98 (T)
cryolite 109
currency area 134

dairy industry 114, 126
dates, production of 78, 94 (T)
desalination 79-80, 120
development
    co-operation for 1-6, 52, 135
    expenditure 82, 99 (T)
    obstacles 4
    policy 51, 101
    strategy 3-4
    technological 5, 103-6
diversification, economic 5-6, 45, 49, 71-3, 75, 133-4, 136
    industrial 104, 131
Dubai 132

Economic Agreement of 8 June 1981 4-6
education 18, 37 (T), 78
electric power 107-8
    cables 119
    equipment 117-18
    generating capacity 117; demand for, by country 128 (T); standardization of 118
    see also transformers
electrolytic refining and smelting 106, 110
electronics industry 116
employment
    in agriculture 77, 90-2 (T)
    see also labour force
energy
    consumption increase 11
    resources 20, 39 (T)
    see also natural gas, oil
engineering 114-15
erosion 22
ethane 47
ethylene 48, 50, 67-8 (T)
Europe
    excess petrochemicals production in 49
    Gulf capital cost differentials with 47
expenditure
    development 82, 99 (T)
    on GDP, by country and demand category 7, 11, 27-8 (T)
exports 11-16, 33-4 (T)
    diversification of 104
    tariffs against 105
    see also oil exports
extraction industries, proportion of GDP 9
Exxon Chemical Company 67

farm
    holdings, smallness of 77
    management 84, 133
fertilizer industries 133
    by country 50, 65 (T)
    production equipment 120
fibres, synthetic 48, 52, 132
financial surpluses 3-4, 8, 11, 15, 43, 131
    investment in petrochemicals 49, 51
fish-meal 113
fisheries 21, 39 (T), 72, 113, 134
food
    imports 72, 82
    processing 103, 111-14; by country 113-14, 126 (T); export-oriented 112; water requirements 112-13
    reserves 83
    supply problems 80-2, 133-4
Food and Agriculture Organization

76
French North Company 51
fruit production 78, 94 (T), 96-7 (T)

gas *see* natural gas
gasoline, natural 48
glass manufacture 133
goats 98 (T)
gold 132
graphite electrodes plants 108
greenhouses 79
Gross Domestic Product (GDP)
    by country and sector 7-8, 10, 23-5 (T), 29-31 (T)
    expenditure on, by country and demand category 7, 11, 27-8 (T)
    per capita 8
    sectoral indexes 7, 10, 26 (T)
growth, economic 1-2, 7-9
    balanced, constraints on 72-3
    rates 10, 101
Gulf Cable and Electrical Industries 119
Gulf Petrochemical Corporation (Bahrain) 65, 68

HDPE 51, 67-8
heat exchangers 120-1

illiteracy 18, 37 (T)
import substitution 113
imports 11-12, 14-16, 33 (T), 35 (T), 115, 135
incomes, *per capita*, rise in 81
India 51
industrial chemicals 106
industrial development policy aims 51
industrial processing 104-5
    see also food processing
industrialization 121, 131
    case for 101-2
    social cost of 101, 103
industry
    capital-intensive 108
    diversification of 104, 131
    energy-intensive 107-8
    low labour-intensive 103, 106
    regional co-operation in 102, 115, 119
    resource-based 102, 104-11; vertical perspective on 107,

131
    water requirements 103
    see also capital goods, chemicals, electric power, electronics, extraction industries, fertilizer industries, food processing, manufacturing, natural gas, oil, petrochemicals
inflation 8
infrastructure 71-2, 135
investment 12, 135-6
    absorptive capacity 3, 11, 72
    agricultural 80
    co-ordination of 73
    corporations 135
    in non-oil sectors 71-2
    in petrochemicals 49, 51
iron and steel production 102, 107-8
    by country and product 122-3 (T)
    direct reduction process 106-8
    electric furnace steel plant 107-8
iron ore 111, 132
irrigation 20-1, 76, 78-9, 82, 133
    areas of, by country 87-9 (T)
    cost of 82
    long-term effects of 80
    systems 80

Japan, petrochemicals production in 49, 67

Kuwait 10, 15, 42, 113-14, 119-20
    Desalination Plants Fabrication Company 120-1
    Industrial Refinery Maintenance and Engineering Company (KREMENCO) 120
    Petrochemicals Industries Corporation 65

labour
    female 19, 38 (T)
    force 18, 38 (T); age distribution 19, 38 (T)
    foreign 17, 19, 38 (T), 49, 81, 136
    low intensity use of 103
    recruitment problems 77-8

skilled, scarcity of 77–8,
       82–3, 104–5
   training 82–3
land areas, by country 87–9 (T)
   see also arable land
LDPE 50–1, 67–9
lemons 78, 94 (T)
livestock 82–3
   feed 113–14, 126, 133
   management 78, 133
   numbers, by country 98 (T)
   potential 76, 79
   production 78–9

manufacturing 102, 104, 109
   growth in 11, 101
market integration 5
marketing 106, 133
melons, production of, by country
       96 (T)
metals, non-ferrous 106
methane 47
methanol 48, 50, 67–9 (T)
   see also natural gas
minerals
   metallic 21, 110–11, 132; see
       also bauxite, copper, iron ore
   non-metallic 21, 133
mining 110, 132
   investment in Arab and Third
       World countries 132–3
Mobil Chemical Company 67
multinational corporations see transnational corporations (TNCs)

naphtha steam-cracking 48
national income, oil consumption
       and 8–9
natural gas 43–4
   as energy input for industry 102,
       107–8, 110
   chemical structure of 59–60 (T)
   extraction and processing equipment 120
   flaring of, in decline 44, 56 (T)
   from oil-less reservoirs 44–5
   liquification 47–8, 131; production capacity 48, 63–4 (T);
       projects, by country and
       product 63–4 (T)
   production 44; and utilization,
       by country 56 (T); by country
       44, 57 (T)
   reserves 20, 41, 44–5; by country
       58 (T), 131
natural resources 20–2, 39–40 (T)
   common to region 72
   for industry 102, 104–5, 107
   of Arab and Third World
       countries 132–3
   regional trade in 132
nuclear power 117

oil
   as energy input for industry 102,
       111
   conservation 41
   exports 16, 42; by country
       55 (T); reduction of, to
       promote refining 47; refined,
       capacity for, by country 61
       (T)
   industry, plans for 43–4, 46
   output 8, 41–2; by country 10,
       54 (T); discretionary 42–3
   prices 8–10, 42, 114
   productivity of wells 42
   refineries 46–7, 62 (T)
   refining 46–7, 106; capacity, by
       country 61 (T); equipment
       120
   reserves 41–2, 53 (T), 131
   resources, life of 1, 41–3, 45, 121
   revenues, social cost of 135
   transport 42
oil-seeds 79
   crushing plants 112
olefins 48, 132
Oman
   agricultural potential 72, 76, 79,
       81, 95 (T)
   industry 106, 110, 113, 114
Organization of Petroleum-Exporting
       Countries (OPEC)
   gas production and reserves, Gulf
       proportion of 44–5
   oil exports, output and reserves,
       Gulf share of 41–2

Pakistan 51
Payments Union 134
petrochemicals 45–52, 131–2
   industries 49–52; by country
       65 (T), 67–9 (T)
   investment in 49, 51
   price-undercutting in 49
   production 49–51; machinery and
       equipment for 120–1; regional

co-operation in 131-2; trade policy and 132
projects: abroad, financing of 51-2; by country and product 67-9 (T); costs 49, 52
petroleum coke 109, 111
phosphates 21, 133
plastics 48
polyethylene 119
polyvinyl chloride 119
population 17
  agricultural 77, 90-2 (T)
  increase 8, 16, 113
  statistics 16-19, 37 (T), 70 (T)
poultry 79, 82, 113, 126
price undercutting in petrochemicals 49
production structure 7-12
  see also Gross Domestic Product
propane 48
propylene 48, 68-9

quartz sand 133

raw materials, importing 132
  see also natural resources
re-exports 13-14
refineries, Gulf proportion of 46
  see also under oil
region, Gulf see co-operation, regional; natural resources; trade, intra-regional
rentier economy 135-6
rubber 48

salt build-up 80
Saudi Arabia 10, 108, 110-11, 118-20, 132-3
  agriculture in 72, 76, 79, 81, 95 (T)
  Basic Industrial Corporation (SABIC) 50
  food processing in 114, 126 (T)
  petrochemical companies 67-8
scale economies 105, 113
sheep 98 (T)
Shell Oil Company 67
shipping 105
solar energy 20
Sri Lanka 51
standards of living 1, 45
steel alloy 108
  see also iron and steel
styrene 50, 67, 69

subsidies
  agricultural 80
  food imports 72
Sudan 82, 85
sulphur 133

tariffs on industrial exports 105
technological development 5, 103-6
technology, joint acquisition of 5
telecommunications equipment 115-17, 119
  demand for 127-8 (T)
telephone
  cables 119
  electronic digital exchanges 116
  equipment, demand for, by country 116, 127 (T)
telex equipment, demand for 116-17 by country 128 (T)
Third World, joint processing complexes in 52, 132
timber 21
tomato products, by country 96 (T)
trade
  commodity 13-15, 32 (T)
  external 11-15, 32-5 (T)
  increase in 11, 15
  intra-regional 2, 12-14, 32 (T), 132
  policy, petrochemicals and 132
transformers
  capacity of, in Gulf region 117-18
  demand for 118, 129 (T)
  standardization of 118
trans-national corporations (TNCs) 48-50, 104-5
  joint ventures with 51-2, 132
transport 5, 134
  costs 47, 105, 112
  of refined oil 46-7
Turkish Mediterranean Petrochemical Company 51

United Arab Emirates (UAE) 76, 119, 132
urea 50, 65-6 (T)

waste disposal 22
waste of wealth 136
water
  desalination 79-80, 120
  for cooling 79
  industrial use of 103, 112-13

management 84, 134
opportunity cost of 82, 112
pricing 134
recycling technology 113

resources 20, 39 (T), 79; common to region 72
women workers 19, 38 (T)